Working Across Boundaries

Working Across Boundaries

Resilient Health Care

Volume 5

Edited by
Jeffrey Braithwaite, Erik Hollnagel, and
Garth S Hunte

CRC Press is an imprint of the
Taylor & Francis Group, an informa business

CRC Press
Taylor & Francis Group
6000 Broken Sound Parkway NW, Suite 300
Boca Raton, FL 33487-2742

Printed on acid-free paper

International Standard Book Number-13: 978-0-367-22457-8 (Paperback)
International Standard Book Number-13: 978-0-367-22459-2 (Hardback)

Library of Congress Cataloging-in-Publication Data

Names: Braithwaite, Jeffrey, 1954- editor.
Title: Working across boundaries. Volume 5, Resilient health care / edited by Jeffrey Braithwaite, Erik Hollnagel, and Garth Hunte.
Description: Boca Raton : Taylor & Francis, 2018. | "A CRC title, part of the Taylor & Francis imprint, a member of the Taylor & Francis Group, the academic division of T&F Informa plc." | Includes bibliographical references.
Identifiers: LCCN 2019002062| ISBN 9780367224578 (pbk. : alk. paper) | ISBN 9780367224592 (hardback : alk. paper) | ISBN 9780429274978 (e-book)
Subjects: LCSH: Health services administration. | Organizational effectivenes. | Interorganizational relations.
Classification: LCC RA971 .W67 2018 | DDC 362.1068—dc23
LC record available at https://lccn.loc.gov/2019002062

Visit the Taylor & Francis Web site at
http://www.taylorandfrancis.com

and the CRC Press Web site at
http://www.crcpress.com

Contents

Part IV Empiricising Boundaries

Part V Closure

Preface

Workplaces and institutional life are growing ever more complex. It is not merely that bureaucracy is on the march, or that technological solutions are appearing more rapidly than before, or that software, which already rules almost everything, keeps adding new features and add-ons at a dizzying pace. It is not just that there are more specialisations and super-specialisations filling in seemingly every organisational and professional crevice. It is not just that consumers are more educated, more demanding, and more discerning. It is not because there are financial and economic pressures everywhere. It is not even that there are jobs that exist today (e.g., social media director, well-being manager, drone technician, genetic counsellor) that were not even imagined 20 years ago. It is not just that people are increasingly unable to anticipate the full consequences of the changes and improvements they feel forced to make to overcome acute organisational problems.

It is all of these, and more – in health care as well as everywhere else. Indeed, health care has been described as the complexity exemplar *par excellence* (Braithwaite et al., 2017). There are hundreds of roles in health systems, not only to provide care to patients, or to support those who provide that care, but also in the plethora of service functions underneath the smooth surface of any efficient organisation. There are dense social and professional structures, complicated ecosystems striving to deliver care to patients with complex conditions, and a multiplicity of types of information, communication, clinical and diagnostic technologies applied to those caring processes.

Central to our own interests in this, the fifth volume of the *Resilient Health Care* series, is patient safety – but in the sense of how care goes well. Keeping patients safe, as we have learned across the pages of the previous four volumes (Hollnagel, Braithwaite, & Wears, 2013; Wears, Hollnagel, & Braithwaite, 2015; Braithwaite, Wears, & Hollnagel, 2016; Hollnagel, Braithwaite and Wears, 2019), is incredibly challenging. To do so, people must constantly adjust to both predictable and unpredictable circumstances. And while providing safe care, they must traverse multiple divides – social, cultural, technological, professional, financial, operational – to achieve their aims. They have to work across boundaries, such as the space between Work-as-Imagined and Work-as-Done, or the gaps in understanding they have to live with, or the silos that all human systems create in every workplace. All boundaries have an intended purpose, but given that it is impossible fully to anticipate all possible consequences, many also have unintended side effects. Likewise, people may also introduce temporary boundaries to be able to manage and protect their own work. People think ahead, recognise patterns, make sense and heedfully interrelate. They try to avoid system brittleness, use slack to provide a buffer, and learn about and do their work as it actually unfolds, rather than in accordance with a policy manual. They tailor the work environment, the system, and their tasks to achieve their purposes. In short, agents in health care operate in adroit ways in turbulent and stretched systems, working around problems, and often deal with things that have never happened before in quite the same way.

For all these reasons, we have titled this book *Working Across Boundaries*. The volume pays homage to and attempts to do justice to this galvanising theme, asking how do people navigate the world of work in health settings? In what ways can providers, patients, planners, developers, researchers, policymakers, managers – indeed, everyone – engage with others to deliver safe, effective, evidence-based care?

These are the thorny questions we began to formulate as we heard the presentations at the Resilient Health Care Network Meeting in Middelfart, Denmark in August 2016, where 60 of us gathered for our annual get together. They represent the challenge we threw at our chapter writers, who responded by figuring out their own answers to the problematic: how do people ensure that work goes well while negotiating the permanent and temporary boundaries they encounter in modern health settings? That was the task; and now we will turn the pages to see how well they fared.

Jeffrey Braithwaite, Erik Hollnagel, and Garth S. Hunte

REFERENCES

Braithwaite, J., Churruca, K., Ellis, L. A., Long, J., Clay-Williams, R., Damen, N., … Ludlow, K. (2017). *Complexity Science in Healthcare – Aspirations, Approaches, Applications and Accomplishments: A White Paper.* Sydney, Australia: Australian Institute of Health Innovation, Macquarie University.

Braithwaite, J., Wears, R. L., & Hollnagel, E. (Eds.). (2016). *Resilient Health Care, Volume 3: Reconciling Work-as-Imagined and Work-as-Done.* Farnham, UK: Taylor & Francis Group.

Hollnagel, E., Braithwaite, J., & Wears, R. (Eds.). (2013). *Resilient Health Care.* Farnham, UK: Ashgate Publishing.

Hollnagel, E., Braithwaite, J., & Wears, R. (Eds.). (2018). *Resilient Health Care, Volume 4: Delivering Resilient Health Care.* Abingdon, Oxon: Routledge.

Wears, R., Hollnagel, E., & Braithwaite, J. (Eds.). (2015). *Resilient Health Care, Volume 2: The Resilience of Everyday Clinical Work.* Farnham, UK: Ashgate Publishing.

Acknowledgements

We dedicate this book to a giant in the field of patient safety, resilient health care and emergency care – our friend, supporter and teacher, Bob Wears. Vale, Bob – we miss you.

Much gratitude goes to our international colleagues, the chapter authors, who have applied ingenuity, expertise and endeavour, which is reflected in every part of the volume. Every chapter is a rigorous yet creative endeavour. The culmination of this book is a testament to the outstanding and diverse range of expertise in the Resilient Health Care Network.

Our thanks also go to the editorial team at the Australian Institute of Health Innovation, Sydney, Australia for their efforts and support in the creation of this fifth instalment of the series. Dr Wendy James copy edited each chapter. Our research assistant team formatted and pulled everything together (Anne Grødahl); created the index and proofed chapters (Claire Boyling) and edited and proofed each individual chapter and sourced the biographies (Meagan Warwick).

JB, EH, GSH
Sydney, Nivå, Vancouver
January 2019

Editors

Jeffrey Braithwaite, BA, MIR (Hons), MBA, DipLR, PhD, FIML, FCHSM, FFPHRCP (UK), FAcSS (UK), Hon FRACMA, FAHMS, is a founding director, Australian Institute of Health Innovation (AIHI); director, Centre for Healthcare Resilience and Implementation Science and professor of Health Systems Research, Faculty of Medicine and Health Sciences, Macquarie University, Australia. His research examines the changing nature of complex health systems, attracting funding of more than AU$111 million (€71 million, £63 million). He has contributed over 450 peer-reviewed publications, including 12 previous books, presented at international and national conferences on more than 930 occasions, including over 90 keynote addresses. His research appeared in journals such as *Journal of the American Medical Association, BMC Medicine, The British Medical Journal, The Lancet, Social Science & Medicine, BMJ Quality & Safety* and *International Journal for Quality in Health Care.* He has received 43 different national and international awards for his teaching and research. Further details are available on the AIHI website: http://aihi.mq.edu.au/people/professor-jeffrey-braithwaite.

Erik Hollnagel, MSc, PhD, is a senior professor of patient safety at Jönköping University (Sweden), visiting professorial fellow at the Centre for Healthcare Resilience and Implementation Science, Macquarie University, Australia, and professor emeritus at the Department of Computer Science, University of Linköping, Sweden. He has through his career worked at universities, research centres and industries in several countries and with problems from many domains including nuclear power generation, aerospace and aviation, software engineering, land-based traffic and health care. His professional interests include industrial safety, resilience engineering, patient safety, accident investigation and modelling large-scale socio-technical systems. He has published widely and is the author or editor of 24 books, including five books on resilience engineering, as well as a large number of papers and book chapters. The latest titles are *Safety-I and Safety-II in Practice* and *Delivering Resilient Health Care.*

Garth S. Hunte, MD, PhD, FCFP, is a clinical professor and emergency physician at St. Paul's Hospital, a scientist at the Centre for Health Evaluation and Outcome Sciences, Providence Health Care Research Institute, and the strategic lead for Patient Safety and System Resilience in Emergency Care in the Department of Emergency Medicine, University of British Columbia in Vancouver, Canada. His research program is centred around how safety is created in complex socio-technical systems, and in the application of resilience engineering in health care. He is actively involved in the Resilience Engineering Association and the Resilient Health Care Network, and was the organiser and host of the 6th Resilient Health Care Meeting in Vancouver in 2017.

Contributors

Janet E. Anderson
Centre for Applied Resilience (CARe)
Florence Nightingale Faculty of Nursing and Midwifery
King's College London
London, United Kingdom

Jonathan Back
Centre for Applied Resilience (CARe)
King's College London
London, United Kingdom

Deborah Biggerstaff
Mental Health and Wellbeing
Warwick Medical School
University of Warwick
Coventry, United Kingdom

Brette Blakely
Australian Institute of Health Innovation
Faculty of Medicine and Health Sciences
Macquarie University
Sydney, Australia

Jeffrey Braithwaite
Australian Institute of Health Innovation
Faculty of Medicine and Health Sciences
Macquarie University
Sydney, Australia

Kate Churruca
Australian Institute of Health Innovation
Faculty of Medicine and Health Sciences
Macquarie University
Sydney, Australia

Robyn Clay-Williams
Australian Institute of Health Innovation
Faculty of Medicine and Health Sciences
Macquarie University
Sydney, Australia

Marianne Hald Clemmensen
The Danish Research Unit for Hospital Pharmacy
Amgros Copenhagen University Hospital
Copenhagen, Denmark

Ellen S. Deutsch
Pennsylvania Patient Safety Authority
Harrisburg, Pennsylvania
and
Anesthesiology and Critical Care Medicine
The Children's Hospital of Philadelphia
Philadelphia, Pennsylvania

Peter Dieckmann
Center for Human Resources
Copenhagen Academy for Medical Education and Simulation (CAMES)
Capital Region of Denmark
Herlev, Denmark

Louise A. Ellis
Australian Institute of Health Innovation
Faculty of Medicine and Health Sciences
Macquarie University
Sydney, Australia

Sudeep Hegde
Human Computer Interaction
University at Buffalo
Boston, Massachusetts

Katherine Henderson
Emergency Medicine
Guy's and St. Thomas's NHS Foundation Trust
London, United Kingdom

Erik Hollnagel
Hälsohögskolan i Jönköping
Jönköping University
Jönköping, Sweden

Garth S. Hunte
Department of Emergency Medicine
Faculty of Medicine
University of British Columbia
and
Centre for Health Evaluation and Outcome Sciences (CHEOS)
Providence Health Care Research Institute
British Columbia, Canada

Huayi Huang
Warwick Medical School
University of Warwick
Coventry, United Kingdom

Cullen Jackson
Department of Anesthesia, Critical Care & Pain Medicine
Harvard Medical School
Beth Israel Deaconess Medical Center
Boston, Massachusetts

Peter Jaye
Emergency Medicine
Guy's and St. Thomas's NHS Foundation Trust
King's Health Partners Academic Health Sciences Centre
London, United Kingdom

Andrew Johnson
Medical Services
Townsville Hospital and Health Service
Queensland, Australia

Michael Klug
Clayton Utz
Queensland, Australia

Saadi Lahlou
Department of Psychological and Behavioural Science
The London School of Economics and Political Science
London, United Kingdom

Paul Lane
Health and Wellbeing Service Group
Townsville Hospital and Health Service
Queensland, Australia

Eric Arne Lofquist
Department of Leadership and Organizational Behaviour
BI Norwegian Business School
Bergen, Norway

Janet C. Long
Australian Institute of Health Innovation
Faculty of Medicine and Health Sciences
Macquarie University
Sydney, Australia

Mary D. Patterson
Department of Emergency Medicine
University of Florida
Jacksonville, Florida
and
Akron Children's Hospital Simulation Center for Safety and Reliability
Akron, Ohio

Rob Robson
Healthcare System Safety and Accountability (HSSA)
Institute for Healthcare Communication
Ottawa, Canada

Alastair J. Ross
Behavioural Science Glasgow Dental School
University of Glasgow
Glasgow, Scotland
and
Centre for Applied Resilience in Healthcare (CARe)
King's College London
London, United Kingdom

Tarcisio Abreu Saurin
Industrial Engineering Department
Federal University of Rio Grande do Sul (UFRGS)
Porto Alegre, Brazil

Siva Senthuran
Intensive Care Medicine
Townsville Hospital and Health Service
Queensland, Australia

Sam Sheps
School of Population and Public Health
University of British Columbia Vancouver
British Columbia, Canada

Mark A. Sujan
Patient Safety
Warwick Medical School
University of Warwick
Coventry, United Kingdom

Marlon Soliman
Industrial Engineering Department
Federal University of Rio Grande do Sul (UFRGS)
Porto Alegre, Brazil

Robert L. Wears
Emergency Medicine
University of Florida
Jacksonville, Florida
and
Clinical Safety Research Unit
Imperial College London
London, United Kingdom

Natália Basso Werle
Industrial Engineering Department
Federal University of Rio Grande do Sul (UFRGS)
Porto Alegre, Brazil

Lev Zhuravsky
Department of Population Health
University of Otago
Christchurch, New Zealand
and
Patients Care and Access
Waitemata District Health Board
Auckland, New Zealand

Part I

Openings

1

Introduction: The Journey to Here and What Happens Next

Jeffrey Braithwaite
Macquarie University

Erik Hollnagel
Jönköping University

Garth S. Hunte
University of British Columbia

CONTENTS

Introduction

Building on the previous four volumes (Hollnagel, Braithwaite, & Wears, 2013, 2018; Wears, Hollnagel, & Braithwaite, 2015; Braithwaite, Wears, & Hollnagel, 2016), this fifth volume in the Resilient Health Care (RHC) series considers ways of working across various boundaries – and across systems, organisations and services – to enable resilient performance. Everywhere you look in health care, there are bounded behaviours, cultures and structures. There are hierarchies and heterarchies, silos between departments, subcultures, political tensions and disagreements about resources and the way they are allocated. There are gulfs between those who do clinical work and those who manage it, and rifts between those who allocate funding and prescribe policy and those who are obliged to carry it out. There are positive aspects to being bounded – social cliques, groups, professions and clusters, to which we give names like 'the doctors', 'ward 1A', 'the division of nursing', 'the people who go bowling together on Tuesday nights', 'the emergency department', 'the weekend people', 'the old guard' and 'the newbies'. Although boundaries are often associated with negative connotations – that they stop things happening or represent barriers to be overcome – in many situations, boundaries also present clear opportunities or benefits. It is useful and often motivating, for example, to be a circumscribed member of a team or group and to identify who the collective 'you' is, and how 'your' people interact over the fence or across the gap, thereby creating a culture of inclusion. It is likewise useful to establish a temporary boundary around a task, for instance, to ensure sterile conditions or to avoid unnecessary interruptions from others (Mesman, 2009). Boundaries, like barriers, can hinder as well as help.

Significant boundaries can also be found on more conceptual levels; between those who think in Safety-I terms and those who have embraced Safety-II, or between those at the sharp end and those at the blunt end of the system, or between people who focus on Work-as-Imagined and those who actually perform Work-as-Done. But there are other boundaries as well; there are limits of human sociological patterns and psychological characteristics, and the limits of structures and silos of departments, wards, units and entire organisations and systems of care.

The full realisation of the potentialities for RHC across settings requires effective communication across organisational, service-level and functional boundaries. And we must traverse the various silos we find, and bridge the multifarious gaps we experience, whether they are physical, social, technical or conceptual.

Conceptualising Boundaries

To expand the idea of working across boundaries, we need to describe more precisely what this entails. What do we mean by boundaries? And if we can identify a boundary, what does this signify? Here are some ideas in a simple mind map (a depiction of the ideas that are related directly to the central 'boundaries' concept, with a branching out of other ideas from that core idea). These are concepts that are associated with the idea of boundaries as we see the word in use (Figure 1.1).

If boundaries represent the edges, limits, frontiers or perimeters of subsystems, cultures or behaviours, then the next question is to ask what is in the between spaces, those gaps mediating between two boundaries? This between space is often liminal, and can offer advantages and disadvantages, prospects and threats. Here's a second mind map of the gaps that can act at various times and points in a system as buffers and delay mechanisms (Figure 1.2).

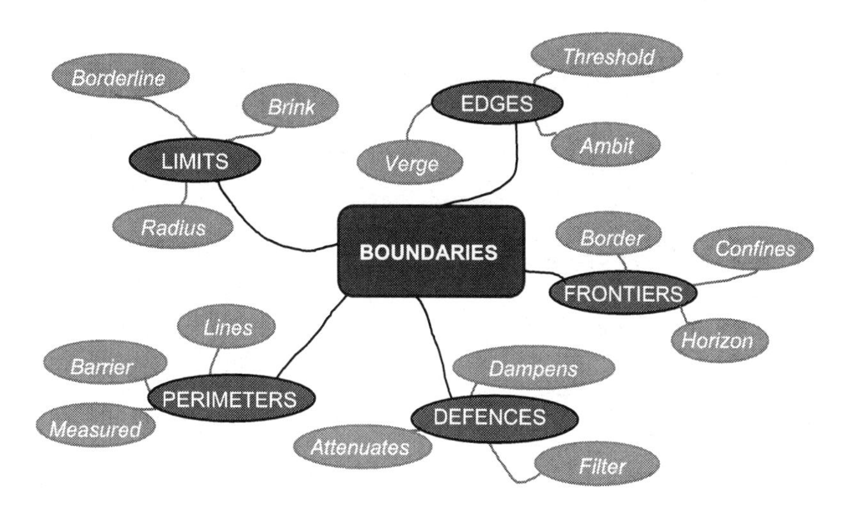

FIGURE 1.1
Boundaries, conceptualised.

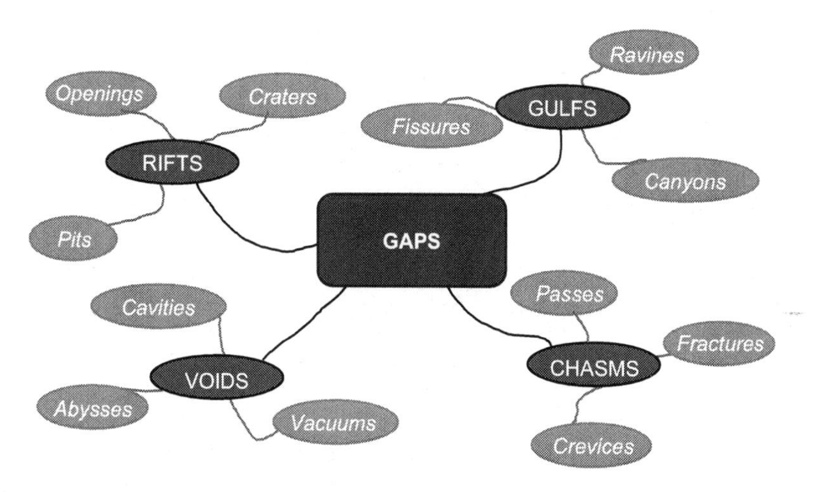

FIGURE 1.2
Gaps, conceptualised.

The third idea, now we have teased out some features of the thesaurus-style relationships of boundaries and gaps, is to ask about gap bridging; a core 'working the boundaries' kind of mechanism. Gap bridging, or action across boundaries, is a way of mediating between gaps separating boundaries. It is inherently vested in people: those who are labelled, for example, as traversers, navigators, cosmopolites or mavens (Figure 1.3).

As Figure 1.3 suggests, gap-bridging activities are often a resource for those who want to stand in the between spaces and traverse them for

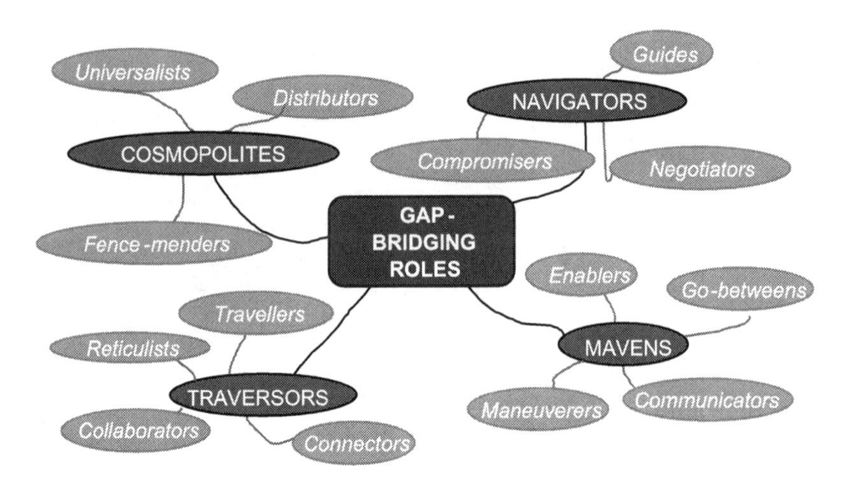

FIGURE 1.3
Gap-bridging roles, conceptualised.

benefit. The people in such a role can be a kind of meso-level activist, or they can be a force for good or a force for ill. Illuminating this is the theory of structural holes in social network analysis, as postulated by Burt (1992). Two main roles according to this theory are *tertius iungens* or the 'third who joins', where holes, gaps or social spaces occur in networks, and this person generates opportunities for others and promotes information flows and creative ideas. The alter ego of this role is the *tertius gaudens,* who creates barriers, plays politics, starves people of information and benefits personally from the role occupied (for further discussion, see Burt, 1992; Braithwaite, 2010, 2015).

Armed with these concepts, we can start to discern what might now be expounded by the authors of the chapters which follow. In parts two, three and four of the book, we will see multiple intellectual and scholarly contributions that tackle this 'working across' idea. We will also learn about the resilient performance of everyday action – about health care activities that are within the boundaries, at the boundaries and in the gaps, along with those that form bridges between them.

The Purpose of the Book

This volume continues the series' journey along the RHC path, arguing the case for recognising and understanding boundaries to use their strengths and overcome their weaknesses, thereby leading to the expression of resilient action in health care. The purpose of *Working Across Boundaries* is to demonstrate how RHC principles can enable those on the front-line to work more effectively – moving towards greater levels of interdisciplinary care, for example – by gaining a deeper understanding of the latent and manifest boundaries that exist in everyday clinical settings. It is also useful for others in health care – patients and advocacy groups, managers, policymakers, patient safety and quality officers – anyone, in fact, who wants to create better health care.

The authors do this by presenting a set of case studies, theoretical chapters and applications that relate experiences, generate ideas and provide practical solutions. As readers will see, the chapters address many different issues, including ways of resolving conflict, overcoming barriers to patient-flow management and building connections through negotiation. The authors have harnessed a range of approaches rather than embracing a single way of solving the practical problems, and their chapters serve both a scientific and an educational purpose.

The Task for Authors: What Happens Next

We invited authors to develop their chapters with two things in mind: first, that our overarching mission in this series is to explore in depth the essence of RHC, and through that, expand our understanding as far as possible. Second, by analysing bridge crossing, manoeuvring, navigating and erecting as well as dissolving barriers, or by simply helping people to apprehend more deeply the boundaries they face and sometimes use, we might enable readers to cope with them, whatever they are and wherever they are found. Thus, we might help produce more joined up and resilient care, or appreciate more clearly how to do so.

That was the goal of this book. In the chapters that follow, we will see how the authors responded to the call to work across boundaries – the directions taken, areas explored and questions answered.

References

Braithwaite, J. (2010). Between-Group Behaviour in Health Care: Gaps, Edges, Boundaries, Disconnections, Weak Ties, Spaces and Holes. A Systematic Review. *BMC Health Services Research, 10*(1), 330.

Braithwaite, J. (2015). Bridging Gaps to Promote Networked Care Between Teams and Groups in Health Delivery Systems: A Systematic Review of Non-Health Literature. *BMJ Open, 5*(9), e006567.

Braithwaite, J., Wears, R. L., & Hollnagel, E. (Eds.). (2016). *Resilient Health Care, Volume 3: Reconciling Work-as-Imagined and Work-as-Done*. Farnham, UK: Taylor & Francis Group.

Burt, R. S. (1992). *Structural Holes: The Social Structure of Competition*. Chicago, IL: University of Illinois at Urbana.

Hollnagel, E., Braithwaite, J., & Wears, R. (Eds.). (2013). *Resilient Health Care*. Farnham, UK: Ashgate Publishing.

Hollnagel, E., Braithwaite, J., & Wears, R. (Eds.). (2018). *Resilient Health Care, Volume 4: Delivering Resilient Health Care*. Abingdon, UK: Routledge.

Mesman, J. (2009). The Geography of Patient Safety: A Topical Analysis of Sterility. *Social Science and Medicine*, *69*(12), 1705–1712.

Wears, R., Hollnagel, E., & Braithwaite, J. (Eds.). (2015). *Resilient Health Care, Volume 2: The Resilience of Everyday Clinical Work*. Farnham, UK: Ashgate Publishing.

2

Bon Voyage: Navigating the Boundaries of Resilient Health Care[1]

Jeffrey Braithwaite
Macquarie University

Erik Hollnagel
Jönköping University

Robert L. Wears
University of Florida
Imperial College London

CONTENTS

It was August 17, 2016; the end of the Resilient Health Care Network's (RHCN's) Summer Meeting. The three of us – Erik, Jeffrey and Bob – boarded Erik's yacht, a Hallberg-Rassy 94, with both sails and an inboard diesel engine, moored at the Middelfart marina. It was one of those long Danish summer evenings, and we were ready for a little downtime after 3 days of intensive discussion.

It was warm weather, a gentle breeze blew, and there wasn't much by way of other craft in that sheltered part of the Baltic Sea known as Little Belt. These weren't challenging nautical conditions by any stretch of imagination. As we sailed, the venue for our meeting – Hindsgavl Slot, a castle dating from the 12th century and now a hotel – was on our starboard side and the privately owned island of Fænø to port, en route to Kolding Fjord.

The RHCN meeting had, as always, been intense, stimulating and productive. Bob once described it as a series of short presentations punctuated by

[1] Within a year of this journey, on July 16, 2017, Bob, our friend, colleague, mentor and muse, died. The gap he leaves is immense, and his absence continues to be felt by all who loved him in the Resilient Health Care Network and around the world of patient safety. We dedicate the rest of the book to him. Vale, Bob.

long discussions: the mirror image of the usual conference, where the typical speaker attempts to fill a passive audience with as much of their wisdom as possible in the short time allotted, with little by way of a question and answer period, or time given over to discussion.

RHCN meetings limit participation to around 60 or so people, and we can't accept every abstract submitted, even from those who are active or long-standing members. Those who do participate rotate between presenting and providing an audience for others. This structure means there's very little drifting off: all are expected to be contributors to the discussions.

RHCN is more cognitively challenging than run-of-the-mill conferences in other ways too. We genuinely, and purposefully, try to dig deep into resilience, striving to understand its nature, contours and dimensions. And the 60 of us don't just participate in the meeting, but eat meals, socialise and network. Some of us exercise or walk about the grounds together. There's rarely a moment throughout the 3 days when we aren't thinking of how we can stretch our minds around how resilience works, what it means and how we can advance our understanding of its potentialities. It's definitely more boot camp than holiday camp, intellectually speaking. Despite or perhaps because of the challenges, all who participate agree: attendance is deeply satisfying.

Right from the RHC's inception, we three have taken on editorial responsibilities in addition to our role as participants. We each make notes throughout the proceedings or during the breaks about the talks we've heard and the discussions they've stimulated, tracking and mapping themes and ideas as they manifest, and working out how best to structure the next book in the RHCN series. Four times previously, the information from the RHCN meeting has provided most of the material for the corresponding book, although we have occasionally looked to supplement it with a chapter or two from others who we know are knowledgeable about the particular theme, but who may not, for some reason or other, have made it to the meeting that year.

Sailing as Resilient Practice

Despite the balmy conditions, sailing that evening wasn't all play. Even as we let the joy of sailing and a tiny bit of salt spray wash over us, we were consolidating our thoughts, and although it was never overt, our combined reflections were helping to shape this book.

Bob had sailed before, so he was able to take the helm on the journey back to shore under the guidance of Erik, who was familiar with the complex channels and potentially dangerous shallows. (It wouldn't be a good look if three people who had staked much of their academic reputations on researching, writing about and understanding safety ran aground on the very afternoon the meeting ended.) Jeffrey, the ultimate landlubber, was made an honorary deckhand – enthusiastically putting his still-remembered Boy Scout knowledge of ropes and knots to good use.

There was much laughter about the way the sailing roles we adopted hadn't been assigned but had emerged organically. Without discussing it, we were being resilient – the three of us flexing and adjusting, pitching in, relishing our roles and doing the tasks required for an effective, smooth and safe journey.

While we were certainly watching out for other boats and shallow channels and were alert to the potential dangers of colliding or running aground as we sailed, we were unconsciously emphasising things going right rather than focusing on eliminating errors. This wasn't a fraught and high-stakes clinical environment, of course, but even so, it occurred to us that we were adopting a natural, light-and-easy, Safety-II approach.

Boundary Navigation

That 2-hour journey was far more fruitful than any formal sit-down meeting we might otherwise have scheduled to develop the structure and contents of this book. We chatted between ourselves as we sailed, then later, after we docked, mused over dinner and generated initial ideas about the book's form. We discussed the talks we'd heard and thought about which would most likely lead to a chapter that could find a logical place in the compendium.

That short sail was as apt a metaphor for this book as we could ever have consciously designed. It may have only been a leisurely cruise around the bay, but the journey exemplified several useful points. We were navigating the course we took, rather than having to tame the existing conditions. Rather than being pre-planned or proceduralised, our behaviours, interactions and contributions emerged, bottom-up. And we were traversing boundaries – role boundaries, physical boundaries, sailing knowledge, navigational boundaries, currents and channels, as well as other boats in the fjord.

Let's freeze the frame on our expedition, which was a pretty, and pretty benign, journey. So, here's a question of substance: how do people navigate the boundaries of health settings, in order to deliver good care? This is a question of some moment, because recurring through the pages of this book and its predecessors in the RHC series (Braithwaite, Wears, & Hollnagel, 2016; Hollnagel, Braithwaite, & Wears, 2013, 2018; Wears, Hollnagel, & Braithwaite, 2015) is the idea that gaps, discontinuities, edges and borders are an inherent and inevitable feature of clinical practice – creating conditions that somehow have to be handled, dealt with, traversed or otherwise managed.

The simple answer is that people navigate health care boundaries all the time, just as we navigated the sailing conditions we encountered.

If by navigating (from the Latin *navigatus,* to sail) we mean to chart a course across a challenging terrain, then people (clinicians, support staff, managers) do this continually in their daily activities. People collaborate, negotiate, exchange, trade or otherwise inter-relate with others in their environments longitudinally, succeeding in delivering care in the face of complex circumstances (Braithwaite, Clay-Williams, Nugus, & Plumb, 2013; Robson, 2015). They do so across all sorts of boundaries – physical, social, tribal, intellectual, ideological, cognitive and political.

We can glimpse this further in chapters that deal with individual agents, groups and teams in situ, and traversing the landscapes of health care. They not only survive and thrive but also keep patients safe, and co-create everyday quality care that matters (Braithwaite, Runciman, & Merry, 2009). To tease this out, and to see the socially textured ways people variously live with, exploit or conquer the boundaries they encounter – and collaborate to execute care that is deeply resilient – is the task to which the rest of the book is devoted.

References

Braithwaite, J., Clay-Williams, R., Nugus, P., & Plumb, J. (2013). Health Care as a Complex Adaptive System. In E. Hollnagel, J. Braithwaite, & R. Wears (Eds.), *Resilient Health Care* (pp. 57–76). Farnham, UK: Ashgate Publishing.

Braithwaite, J., Runciman, W. B., & Merry, A. F. (2009). Towards Safer, Better Healthcare: Harnessing the Natural Properties of Complex Sociotechnical Systems. *Quality & Safety in Health Care,* 18(1), 37–41.

Braithwaite, J., Wears, R. L., & Hollnagel, E. (Eds.). (2016). *Resilient Health Care, Volume 3: Reconciling Work-as-Imagined and Work-as-Done.* Farnham, UK: Taylor & Francis Group.

Hollnagel, E., Braithwaite, J., & Wears, R. (Eds.). (2013). *Resilient Health Care, Volume 1.* Farnham, UK: Ashgate Publishing.

Hollnagel, E., Braithwaite, J., & Wears, R. (Eds.). (2018). *Delivering Resilient Health Care, Volume 4.* New York, NY: Routledge.

Robson, R. (2015). ECW in Complex Adaptive Systems. In R. Wears, E. Hollnagel, & J. Braithwaite (Eds.), *Resilient Health Care, Volume 2: The Resilience of Everyday Clinical Work* (pp. 177–188). Farnham, UK: Ashgate Publishing.

Wears, R., Hollnagel, E., & Braithwaite, J. (Eds.). (2015). *Resilient Health Care, Volume 2: The Resilience of Everyday Clinical Work.* Farnham, UK: Ashgate Publishing.

Part II

Negotiating Across Boundaries

3

Working across Boundaries: Creating Value and Producing Safety in Health Care Using Empathic Negotiation Skills

Andrew Johnson and Paul Lane
Townsville Hospital and Health Service

Michael Klug
Clayton Utz

Robyn Clay-Williams
Macquarie University

CONTENTS

Background

Negotiation is a process whereby two or more people resolve or navigate through their differences when there is ambiguity as to the correct outcome. Interest-Based Bargaining (IBB) is a form of negotiation where parties resolve differences by understanding and aligning the interests of the other party to their own. At the core of negotiation theory is the differentiation between integrative (interest-based) bargaining and distributive bargaining. This creates the possibility of creating, rather than merely distributing value for the parties, the classic win-win solution.

A by-product of this process is the development of respect for the other party by understanding more about their interests and goals, thereby working across boundaries and building connections and thus creating value.

While the theories and practice of negotiation have been well described in the management literature, uptake of mainstream negotiation theory within health care has been limited. The fundamental activity of health care funders and providers is to influence behaviour and provide services to create better health outcomes for the community. This is best achieved by the application of negotiation skills to establish what is required and how best to provide it. The relevance of negotiation to clinical practice is not as well understood, but is possibly as important as other clinical skills. It is a critical mechanism to influence behaviour and build rapport with patients, co-workers and stakeholders.

There is, however, an inherent contradiction in the delivery of health services. A consequence of the normal outcome-based objectives – improved health, emergency services, surgical interventions and so on – can be to encourage distributive or competitive behaviour, often in circumstances of urgency. This can lead to 'the end justifies the means' behaviour, which focuses on outcomes to the detriment of relationships.

There is a cultural and cost-saving bounty to be harvested by learning true competency in the negotiation space whilst understanding that not only does this not involve sacrificing quality of outcomes, it enhances it measurably. Value can be created at low or no cost to the health care system by teaching this often-ignored discipline, without sacrificing professional expertise.

The Case for Negotiation Skills in Health Care

In 2014, the Queensland public health system was involved in a protracted and damaging dispute with senior medical staff around employment contracts. A pilot program for negotiation skills training was developed by a leading negotiation skills expert in a commercial arrangement with the Department of Health. The outstanding feedback from this program has led to a broader application within our health service, in terms of training for senior clinicians and health care managers. The training was formally evaluated in 2015 (Clay-Williams et al., 2018), and the resulting changes in practice have led to significant initiatives for sustained improvement.

Malhotra and Malhotra (2013) write about negotiation as a skill to apply in clinical practice and offer cogent examples of how negotiation can influence patients, other practitioners and senior management to achieve better outcomes. We found similar examples from our training. Anastakis (2003) describes the Needs-Based model of negotiation and applies this as a key skill for practitioners to manage their departments and to access the resources that they require from management.

The work undertaken on understanding health care as a Complex Adaptive System (CAS) (Braithwaite, Clay-Williams, Nugus, & Plumb, 2013; Johnson &

Lane, 2016) helps to explain why negotiation may assist in achieving outcomes through influence rather than application of authority. The reasoning behind why conflict is more likely to arise in a CAS is given by Robson in a later chapter. Lane, Clay-Williams, and Johnson (2015) have previously described the TenC model for effective working in a CAS.

This model proposes ten key inter-related qualities of agents within a health care system that come together to create resilience within the system and is currently undergoing validation. Each element (agent or quality) has a consequence on the others and each may compensate for the others. The Ten C's are as follows: Cohesion; Capture; Cognition; Communication; Culture; Clear Ownership; Constraints; Challenge; Competence; and Compliance.

Unlike the typical theoretical based models seen in health care, the TenC concepts have emerged over the past decade from the experience and reflection of front-line hospital clinicians and managers. Negotiation skills are proposed as a principal mechanism for establishing Cohesion in the workplace (always a contested idea, but the key ingredient in the model).

Since the development of the TenC Model, Johnson, Clay-Williams and Lane have developed a Framework for Better Care (Johnson, Clay-Williams, & Lane, 2017) which identifies negotiation skills as a core tool in navigating to effective care in situations of irreducible complexity and uncertainty of outcomes. This framework has been adopted by the Royal Australasian College of Medical Administrators (RACMA) to explain the systems thinking, underpinning their new Clinical Governance Framework (Clay-Williams, Travaglia, Hibbert, & Braithwaite, 2017). Recently, in their Institute of Healthcare Improvement 2017 White Paper, Frankel, Haraden, Federico, and Lenoci-Edwards (2017) identified negotiation as one of the nine elements required for safe, reliable and effective care.

Negotiation Practice in Health Care

There are many different approaches to negotiation available in the academic literature and popular press. These approaches, which are discussed more fully by Robson in his chapter in this volume, generally fall into categories of power-based, rule-based, interest-based and social constructivist views. Each has its critics and, arguably, each has its place. In this chapter, we explore the applications of an interest-based approach.

Health care is the most human of endeavours. It is a world full of people and their interests, where there is a huge value to relationships in achieving outcomes for people, patients and staff; organisations and communities. In this chapter, we focus on one theory and practice of negotiation, IBB (also known as 'principled bargaining'; Fisher, Ury, & Patton, 2011). Whilst variously criticised for trivialising conflict, underplaying the role of context and situation and over-simplifying methods of negotiation, we believe that this approach offers great promise in application, in a nuanced form, in health care.

Our premise, then, is this. To the extent that boundaries exist in the fluid environment of a CAS, they are not static, they are not impermeable, they shift with time and they are inherently amenable to negotiation.

Interest-Based (Integrative) Bargaining

A useful starting point is the distinction between IBB (integrative) and distributive bargaining. In health care, as in other endeavours, we tend to see negotiations as a mechanism to 'get what we want', without exploring the true basis for why we want it. This inevitably lands us in distributive bargaining, where a good outcome for one party may diminish the outcome for the other. In integrative bargaining, the key is to understand *why* rather than *what*. Malhotra and Malhotra (2013) invites us to focus on interests that may be compatible, rather than outcomes that may be irreconcilable and mutually exclusive. In this chapter, we explore the application of integrative bargaining in health care and show how it offers a key to working across boundaries and improving care.

Two Sisters, One Orange

The classic parable was described by Mary Parker Follet, (Follett, 1940), the founder of IBB in the 1920s, who told the story of two sisters squabbling over the last orange in the pantry. Both sisters argued for the orange and a compromise was reached whereby, in the traditional distributive fashion, the sisters took half of the orange each. Both achieved exactly half of their desired outcome for their intended purpose: one to extract the juice and the other to use the peel to flavour a cake. After they had discarded the remains, the sisters came to the realisation that both could have had 100% of their desired outcome if they had understood the interests of each other.

Courtesy of Andrew Johnson

On reflection, each sister needed to be curious and ask the question. "Why does my sister want the orange?" This curiosity is essential to IBB.

Curiosity is the foundation of another fundamental quality for patient care. According to the former Dean of Students at the University of California, Davis School of Medicine, curiosity is the cornerstone of empathy:

> I believe that it is curiosity that converts strangers (the objects of analysis) into people we can empathize with. To participate in the feelings and ideas of one's patients—to empathize—one must be curious enough to know the patients: their characters, cultures, spiritual and physical responses, hopes, past, and social surrounds.
>
> **Fitzgerald (1999)**

Integrative bargaining could perhaps be described as 'Empathic Negotiation'. We believe empathy to be a key facet for effective leadership and management. It allows us to see the perspectives of others, and in doing so, allows us to explore the possibility of integrating interests and developing effective trade-offs where there is divergence. It allows us to align interests and grow value. Identifying the tension between empathy and assertiveness is a major challenge in negotiation and never more so than in the medical profession, with its sad but often necessary duty to deliver bad news.

CASE STUDY ONE – DEPARTMENT LEVEL, INTENSIVE CARE UNIT ESCALATION PROCEDURE

Intensive Care Unit (ICU) resources are precious and expensive and whilst allocation of available resources can be fraught, interest-based negotiation may provide workable solutions.

A common scenario is two patients competing for one available bed. An example would be two post-operative neurosurgical patients and only one available ICU bed.

The distributive approach would be to deny one patient a bed, which in this case would mean cancelling surgery and the potential for patient harm. An interest-based approach would attempt to understand the issues surrounding the bed pressure and the surgeries being planned.

In an example of this challenge, the order of the operating list was changed such that the shorter, less complex surgery was performed first. The patient was given a prolonged recovery in ICU with close observation whilst the second longer, more difficult case was completed.

When the second patient was ready for ICU admission, the first case could be safely discharged to the neurosurgical ward. By employing an interest-based collaborative negotiation style, the clinical teams could reduce potential harm and grow value for health care delivery without increasing costs.

Styles of Negotiation

Whilst empathy can help us identify the interests of the parties to a negotiation, it is through our various negotiation styles that we then prosecute the actual negotiation. Thomas (1976) is credited by Institute of Health Improvement (IHI) with detailing the five ethical styles of negotiation: Collaboration; Competition; Avoidance; Accommodation and Compromise. Two unethical or inappropriate styles can also be described: Con and Rob.

One of the first tasks in successful negotiation is for the negotiator to understand their own default negotiating style; an effective negotiator will be able to move between different styles, depending on the situation. The tool used in our training to help learners to identify their default style is the commercially available 'Flex Style of Negotiation' developed by Hiam (Lewicki & Hiam, 2007; Lewicki, Hiam, & Olander, 1996). The negotiating approach reflects the degree to which the negotiator applies importance to the outcome of negotiation versus the relationship of participants (Figure 3.1).

The default styles are defined by three main variables. These are the degree to which the participant engages or avoids, accepts rules or redefines them, and gives or takes.

The Collaborator is easily recognised. Collaborators are the people who talk to the cab driver; they face into the elevator and engage in conversation; they want the relationship and the outcome. Meetings with collaborators are enjoyable. They give more than they take, they engage rather than withdraw, and they have limited tolerance for rules. Collaborators often don't understand that others don't work the way they do, and, under stress, may start to compete or move to avoid.

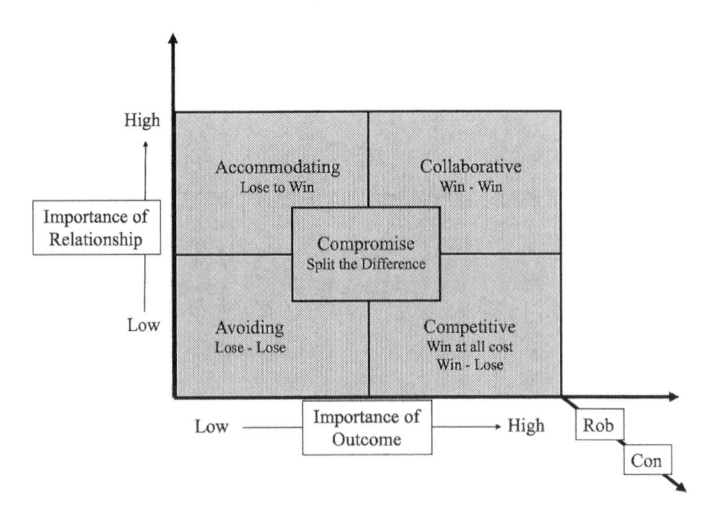

FIGURE 3.1
Negotiation styles. (Adapted from Lewicki and Hiam (2007); Lewicki et al. (1996).)

The Competitor displays an apparent lack of concern for the human dynamic, and they are very much result-oriented. Competitors take, engage and use the rules to their advantage. As they want to win at any cost, they do not develop and nurture long-term relationships that may help create value in the longer term. They justify being purely outcome driven, because they believe they are there 'to get results and not to be nice to people'.

The Avoider just wants to be left alone, 'I'll come to work, I'll do my job and I'll go home'. Avoiders may well be a disgruntled Collaborator who has disengaged or just simply a person who likes to solve complex problems in a solitary manner. Avoiders tend to give more than they take and to withdraw; they tend to think outside the square and have an exceptional eye for detail. Avoiders lose on both the relationship and outcome, but paradoxically, the primary difference between Avoiders and Collaborators is the degree to which they engage.

The Accommodator is often seen as the nicest person in the room. They remember birthdays, kids' names and organise the flowers for the colleagues in distress. They give, engage and accept the rules. Accommodators put great store in relationships and may see the relationship as the desired outcome. The dark side of the Accommodator is that they will tend to give, give and give ... until they start to feel put upon, become resentful and suddenly depart the scene. Others are left wondering what happened.

The Compromiser cuts to the chase. They tap the table at the start of the meeting 'A quick meeting is a good meeting...'. 'Right, are we all done here?' Compromisers value outcomes and relationships, but they are more transactional than Collaborators and will take some gains and move on whilst maintaining a relationship. They take more than they give, tend not to engage at an emotional level and accept the rules. They are quite different from the Collaborator in these respects. Yet natural Collaborators under stress need to learn to flex to compromise, becoming a little more self-protective and less self-destructive than alternative stress-induced responses.

Fortunately, the unethical styles of negotiation, Rob and Con, are unusual in health care. They do exist, however, and working with such individuals can be extraordinarily difficult. Both unethical styles take more than they give, can engage (Con) or withdraw (Rob) and bend or ignore the rules to meet their own needs. It is our experience that these styles can destroy teams, particularly those containing unsuspecting collaborators.

Negotiating with those displaying unethical styles is difficult and requires a different approach. The first action is to look for these styles and recognise that the lack of successful outcome is not the fault of a negotiator using an ethical style. Second, we have found that forming a coalition to tackle the problem is vital; surrounding the party with others who recognise what they are doing. The third is to inform them that you and others recognise the style of negotiation being used and that strong rules, practices and procedures must be employed to improve the situation. This can be difficult for Collaborators, who feel encumbered by rules.

Controlling the flow of information or sphere of influence can also be beneficial. It is important not to back the party into a corner, recognising that giving them an escape route might be better than a lengthy bitter dispute. When given the chance to move on, this party will often find a role where their brilliance and exceptional skill will be clearly recognised.

The authors recognise the difficulties of these negotiations and advise patience and persistence with a rule-based approach. The following chapter on conflict resolution will help provide guidance to manage the situations that often arise with people behaving with these styles.

Of course, none of these negotiation styles are black and white, and there are gradations in every dimension. However, understanding your default position, and what happens to your negotiation style under stress is key to working with others to create, rather than distribute, value.

The IHI White Paper (Frankel et al., 2017) suggests that 'Health care teams should commit to using collaborative negotiation whenever possible. This is the **only** negotiation approach that yields workable solutions that manage resources, provide the best options for patients, and preserve the relationships between parties' (p. 16).

Whilst it is generally true that collaboration has the best chance of creating value, it is not true to say that it is the only approach that yields workable solutions. The most effective negotiators ably flex their style to meet the needs of the situation. At times, this means that a choice can be made to accommodate and concede a point when the outcome doesn't matter as much as preserving the long-term relationship; at other times, competition may be required, and relationships may have to take second place to get a critical outcome in a tight situation. Within a negotiation process, there is usually a place for 'concessions', that is creating a trade-off of one aspect of a negotiation for another that is of greater importance. By its very nature, this process shifts boundaries, establishing opportunities for mutual advantage.

Whilst recognising your own default style of negotiation sits at the core of effectiveness in prosecuting negotiations, it is also important to identify not only the interests of other parties, but also their negotiating style. The interplay between your style and that of other parties provides an opportunity for reflection, and with careful planning, an opportunity to help them (or you) flex, if required, to a more constructive, value-creating approach. That is, helping them to recognise the style they are displaying and encouraging them to flex, just as you do so for yourself.

'Green Credits'

'Green Credits' is a construct adapted from carbon trading and developed by one of the authors (MK), a recognised expert in the field of negotiation and alternative dispute resolution. It denotes the accumulation of trust over the period of a negotiation process. Negotiation can be thought of as a transactional, case-by-case activity, or it can be recognised as part of a longer-term process, where the experience of one negotiation will inform the next. In this latter context, it is possible to accrue 'green credits', where application of an ethical and value-creating negotiating style creates a relationship of sufficient value that the other party recognises the need on occasions to concede a point to you or for you.

That is, a trusted colleague with whom there is a long-standing relationship of fairness and reciprocity which has been built through countless negotiations, is more likely to 'forgive' a competitive negotiation, perhaps allowing the negotiator to draw down a little on a 'bank of green credits'. This is done in the knowledge of the importance of the relationship and the likely reciprocity into the future. It follows that to maintain this level of trust and forgiveness, the negotiator should never 'trade in deficit'. That is, they should consciously consider the state of their relationship credits, when deciding whether to draw down on that relationship in difficult circumstances.

Fisher, Ury and Patton (2011) provides excellent guidance in the development of 'green credits' with the unconditional constructive strategy. This approach allows the negotiator to always act with good intention without compromising their position.

'Green credits' are bankable, transferable and can even be cross-generational. They accrue not only to individuals but also to organisations and institutions (teaching hospitals, universities and charitable organisations).

CASE STUDY TWO – SYSTEM-LEVEL MEDICAL IBB

In 2005, the Queensland health care system was in turmoil. After years of industrial neglect, the State was an uncompetitive employer. Attraction and retention of quality practitioners, particularly in regional areas, were particularly problematic under an award-based industrial agreement.

Following a crisis precipitated by the 'Dr Death' scandal in the regional city of Bundaberg, the State government authorised the use of an alternative form of industrial negotiation to address the interests of the doctors, whilst endeavouring to create value for the system and the stakeholders.

Over several months, one of the authors (MK) facilitated a series of negotiations with parties to the industrial process: unions, practitioner organisations, employers and state department officials. The process followed the principles of IBB and was described as Medical IBB.

Through this process, each of the parties identified the interests of their group. There was a substantial overlap. This allowed for exploration of how to integrate the remaining interests and develop trade-offs.

For example, in hours of work: more flexibility in hours for the practitioner, full-time (40 hours) in 4 days rather than 5; and more flexibility for the employer in the ability to roster 'normal hours' of work beyond the previously accepted boundaries of 0800–1800.

The result was an environment to create a more flexible and attractive award-based employment arrangements in Australia, attractive for the practitioner and effective for the employer.

Conclusion

The rapport of the doctor–patient relationship has at its heart a negotiating style based upon respect and understanding each other's interests. Given the increasing complexity within health care delivery and the increasing variety of treatment options available, there has never been a more important moment for developing robust negotiation skills for clinicians. Only by fully understanding the interests of patients and their families that the best treatment decision be made.

In an era where health care costs continue to rise, creating financial burdens for patients, clinicians, managers and executive, the role of interest-based negotiation – growing value without increasing costs through endeavouring to achieve a win-win outcome (always aspirational, but not always achievable) – has never been more important. It is our experience that this happens best with planning, and by using tools to assist the negotiation process.

In a health care system of silos, where specialist clinicians must be brought together – often by a perplexed medical leader – to solve a complex problem, the interest-based negotiation process offers a framework for completing this difficult task. Empathy and respect are often gained through understanding the issues of the other party, and greater cooperation between various patient teams improves patient care. Negotiation offers the key to working across boundaries, creating value and producing safety in health care.

References

Anastakis, D. J. (2003). Negotiation Skills for Physicians. *The American Journal of Surgery, 185*(1), 74–78.

Braithwaite, J., Clay-Williams, R., Nugus, P., & Plumb, J. (2013). Health Care as a Complex Adaptive System. In E. Hollnagel, J. Braithwaite, & R. Wears (Eds.), *Resilient Health Care* (pp. 57–73). Farnham, UK: Ashgate Publishing.

Clay-Williams, R., Johnson, A., Lane, P., Li, Z., Camilleri, L., Winata, T., & Klug, M. (2018). Collaboration in a Competitive Healthcare System: Negotiation 101 for Clinicians. *Journal of Health Organization and Management, 32*(2), 263–278.

Clay-Williams, R., Travaglia, J., Hibbert, P., & Braithwaite, J. (2017). *Clinical Governance Framework: A Report Prepared for the Royal Australasian College of Medical Administrators (RACMA).* Melbourne, Australia: RACMA.

Fisher, R., Ury, W. L., & Patton, B. (2011). *Getting to Yes: Negotiating Agreement Without Giving In.* New York, NY: Penguin Books.

Fitzgerald, F. (1999). On Being a Doctor. *Annals of Internal Medicine, 130*(1), 70.

Follett, M. (1940). Constructive Conflict. In H. Metcalf & L. Urwick (Eds.), *Dynamic Administration: The Collected Papers of Mary Parker Follett* (pp. 30–49). New York, NY: Harper.

Frankel, A., Haraden, C., Federico, F., & Lenoci-Edwards, J. A. (2017). *Framework for Safe, Reliable, and Effective Care: A White Paper.* Cambridge, MA: Institute for Healthcare Improvement and Safe & Reliable Healthcare.

Johnson, A., Clay-Williams, R., & Lane, P. (2017, August). *Using the Right Tool for the Job in a Resilient Healthcare System – Creating Value through a Framework for Better Care.* Paper presented at the 6th Resilient Health Care International Symposium. Vancouver, Canada.

Johnson, A. & Lane, P. (2016). Resilience 'Work as Done' in Everyday Clinical Work. In J. Braithwaite, R. Wears, & E. Hollnagel (Eds.), *Resilient Health Care, Volume 3: Reconciling Work-as-Imagined and Work-as-Done* (pp. 71–88). Farnham, UK: Ashgate Publishing.

Lane, P., Clay-Williams, R., & Johnson, A. (2015, August). *The TenCs Model: A Case from Townsville Australia.* Paper presented at the 4th Resilient Health Care International Symposium. Sydney, Australia.

Lewicki, R. J. & Hiam, A. (2007). The Flexibility of the Master Negotiator. *Global Business and Organizational Excellence, 26*(2), 25–36.

Lewicki, R. J., Hiam, A., & Olander, K. W. (1996). *Think Before You Speak: A Complete Guide to Strategic Negotiation.* New York, NY: John Wiley & Sons.

Malhotra, D. & Malhotra, M. (2013, October 21). Negotiation Strategies for Doctors – and Hospitals. *Harvard Business Review.*

Thomas, K. W. (1976). Conflict and Conflict Management. In E. A. Locke & M. D. Dunnette (Eds.), *Handbook of Industrial and Organizational Psychology* (pp. 889–935). Chicago, IL: Rand-McNally.

4

Untangling Conflict in Health Care

Rob Robson

Institute for Healthcare Communication

CONTENTS

Background

During a recent vacation in Greece, a well-respected senior health care leader decided to make a side trip to Delphi to seek advice about a problem that had been vexing her for some years. She was hopeful that the oracle at Delphi could provide some guidance. The problem went something like this:

> Despite the fact that our health care facilities and systems are blessed with intelligent, dedicated, hardworking clinicians, caregivers, and managers, we still cannot seem to reach a point where collaborative activity comes naturally – the many silos and mini-kingdoms seem to operate quasi-independently, often to the detriment of safe quality care that responds to the needs of the patients, families and the community. Why does this continue? What is to be done?

After arriving at Delphi and submitting her vexatious problem, the leader was prepared to wait, thinking that this problem might overwhelm the abilities of the oracle. She decided to visit the surrounding countryside, which was truly stunning. When she returned later in the same afternoon, she found the oracle had already produced some (conveniently bullet-pointed) advice:

- It is well known that complex adaptive systems (CASs) generate significant uncertainty and unpredictability; what is less well known is their propensity to create conflict.

- Health care is an exemplar of a CAS, whether this is examined at the level of individual teams, departments, facilities or the system as a whole.
- Conflict will manifest differently at various levels of the system.
- Efforts to address conflict in a given CAS should ideally adopt approaches and techniques that are constant with the characteristics of that system.

The leader was both satisfied and somewhat perplexed. Satisfied because the oracle had provided clear direction, but perplexed as to how the advice could be applied in the context of her specific system. This chapter will deconstruct the advice to determine the most effective and appropriate way to untangle conflict in health care.

CASs and Conflict

What are the important characteristics of a CAS and how do these contribute to conflict? These questions have been reviewed in Volume 2 of *Resilient Health Care* (Robson, 2015) in addition to a recent White Paper from the Australian Institute of Health Innovation (Braithwaite et al., 2017). These resources provide more detail about the evolution of the academic and practical understanding of CASs and, in particular, the way in which conflict emerges as an inherent element of these systems.

Briefly, CASs are typically open systems with flexible and semi-permeable boundaries. This means that the CAS is susceptible to influences from outside the usual limits of the system. Ultimately, a CAS will develop (through a process known as self-organisation) as a result of dynamical, often unpredictable interactions (often occurring simultaneously at multiple levels) between the components and agents that comprise the system at any given moment. These dynamical interactions (the agents are often described as being semi-autonomous, while they primarily interact at local levels within a system, they are influenced by and respond to factors both within and outside the system) help us to understand the many changing properties of the system, by producing patterns of activity through a process known as emergence. Essentially, CASs are fundamentally relational, with the relations between components and semi-autonomous agents defining the nature of a given CAS.

Inherent in a CAS is the concept of uncertainty – sometimes referred to as intractability, or the inability to fully describe all the components, agents and their interactions – which leads to the unpredictability of these systems. Sounds like health care? It is not surprising, given the dynamical nature noted earlier, that a CAS can never be fully planned or controlled. Fortunately, it is possible to learn how to influence self-organisation and emergence with a view to encouraging patterns that are more consistent with the values and goals of the CAS. This is where the question of using appropriate means of addressing conflict within a CAS becomes so important. If we use approaches or techniques that do not reflect the underlying 'operating system' of the CAS, we are unlikely to be able to influence the self-organisation process.

Where does conflict in a CAS originate? With multiple (and multi-level) factors interacting in a dynamical manner, it is inevitable that resources will often be inadequate for the functional operations of some components or subsystems. In health care, this translates most easily into the needs of various domains, units, departments or programs that can never be fully satisfied. It is therefore not a mystery as to why silos develop in health care, competing for insufficient resources while responding (primarily at the local level) to the activities of diverse internal and external agents. It now becomes clear why the frustrated senior health care leader who visited Delphi asked about the 'silos and mini-kingdoms [that] seem to operate quasi-independently'. She was describing (perhaps without fully realising it) an important manifestation of conflict that develops in CAS.

Addressing Conflict and Disputes

Much has been written about conflict, its origins and the variety of approaches to address and resolve it through the active engagement of main players. Foundational texts (Bowling & Hoffman, 2003; Bush & Folger, 1994; Fisher, Ury, & Patton, 1983; Kritek, 2002; Marcus, Dorn, & McNully, 1995; Mayer, 2000; Moore, 1996; Stone, Patton, & Heen, 1999; Ury, 1999) include both the historical understanding of conflict as well as an analysis of the main approaches to dealing with disputes. For a long time, the approaches to conflict were considered to fall into three categories, relying on the application of power, or of rules, or on an appeal to the interests of the disputing parties as preferred methods to resolve issues.

Power-based approaches are exemplified by a number of contemporaneous international situations. The approach is relatively common in organisations and systems that are typically rigid hierarchies, where management by 'command-and-control' is the order of the day. Relying on the power to resolve disputes is also the foundational component of bullying. The resolution of any dispute through a power-based approach is rarely satisfactory and even more rarely permanent. Ury (1999) has written persuasively about this. Of interest, the power-based approach is often softened in CASs when combined with rules, written procedures and codes of conduct.

Rules-based approaches are common in large social organisations (Olson & Eoyang, 2001), and are also a common feature in systems with a command and control hierarchical structure. In these structures, such approaches appear as both implicit and explicit rules, extensive procedure manuals intended to describe in detail how tasks are to be accomplished – often described as 'Work-as-Designed' (Hollnagel, 2015) – in addition to Codes of Conduct. These are the stock-in-trade of many Human Resources (HR) professionals in large organisations. HR professionals are frequently called on to intervene in situations where there is a perceived conflict, alleged inappropriate behaviour or an unanticipated outcome of a work process based on an assumption of 'human error'.

In such circumstances, established rules, procedures or codes of conduct provide a convenient way of apparently 'adjudicating' situations that are felt to be negative or undesirable. Given the way in which CASs evolve and develop, this kind of resolution of conflict can rarely facilitate more productive interactions between agents and components in the system. A rules-based approach provides satisfactory 'accountability outcomes' (clearly identifying who is 'at fault') from the point of view of senior leadership of the organisation.

Interest-based approaches have become the methods most commonly adopted by conflict resolution practitioners and mediators over the last several decades. Most organisations also maintain a rules-based approach that may be combined with an interest-based approach. Most mediators and conflict resolution practitioners associate this approach with the work of Roger Fisher and the Harvard Law School Program on Negotiation (Fisher et al., 1983). Other prominent organisations such as Collaborative Decision Resources Associates in Boulder Colorado have contributed to and broadened the theoretical basis of this approach (Mayer, 2000; Moore, 1996). The underlying assumption of an interest-based approach is that lying behind formal or public positions adopted by the parties to a conflict there may be more fundamental interests that can be shared and explored. This can encourage a search for mutual common interests (or at least overlapping interests) and lead to a potential resolution (the widely hoped for 'win-win' solution).

The role of the mediator in such disputes is to assist the parties to listen and then hear and acknowledge what each is saying as a way of helping them find mutual interests as a pathway to resolution. The concept of the mediator as a neutral party is central to this 'problem-solving' process, and the underlying assumption is that once the various individual interests have been identified a solution can be constructed by the parties to 'unblock' the conflict. This is a fundamentally linear way of viewing the conflict in that the 'problem to be solved' is thought to reside in the positions adopted by the parties. These assumptions make sense in simpler situations and have been successful to a certain extent.

A New View

As the complexity of social organisations has increased, challenges have arisen, leading to a 'social constructionist' analysis of conflict. Some of the philosophical underpinnings of interest-based approaches have been examined in depth and have been found to be less applicable to a CAS. The application of a social constructionist viewpoint has led to the emergence of a relational or narrative approach to addressing conflict in complex situations. An expanding body of writing and reflection (Mayer, 2004; Winslade & Monk, 2000, 2008) has identified multiple factors, which suggest that a relational or narrative approach to conflict is better suited to the particular nature of a CAS. As will be seen, this approach is also better suited to addressing conflict in health care.

A social constructionist perspective identifies a number of problematic issues concerning the use of interest-based approaches in CASs. The first of these is the linear problem-solving framework that is inherent in this approach. The case study offered by Johnson et al. (see Chapter 3), drawn from the experience of the Townsville group, is an excellent example of a conflict involving processes that are both 'tightly coupled' and 'linearly interactive' (Perrow, 1984), where interest-based negotiation approaches can lead to successful outcomes in health care. However, it is important to note that most situations resulting in conflict in a CAS are the result of the dynamical interactions and are much less amenable to influence using a linear approach.

The problem-solving approach that is implicit in interest-based mediation leads to a mindset that perceives conflict as something to be 'fixed' (once this is accomplished, the system will once again run smoothly), very much like the Newtonian or mechanistic view of the world that encourages us to think that we need only find the 'broken parts' to heal a broken process (Capra & Luisi, 2014).

A second important issue concerns the implicit assumption that most conflict originates from within the individuals involved in a dispute – a result of unsatisfied needs that lead to positions that tend to be binary ('my way or the highway') – rather than emerging from the contextual influences in the system itself. Of course, this ignores the fact that the conflicts or disputes reflect a pattern of behaviour that has emerged from the dynamical interactions of multiple factors and components floating – sometimes swimming and sometimes almost drowning – in a sea of contextual influences.

A third central issue that has long been debated within the conflict resolution practitioner community is the assumption of neutrality on the part of the mediator or person who is facilitating the resolution of a dispute. The romantic notion that mediators can somehow rise above their own history and values and remain stalwartly disinterested and neutral with respect to the participants and issues in dispute has been elevated to the legendary status of creed. It does not reflect reality, and this becomes apparent when adopting a narrative approach. The 'story' that makes sense of a particular conflict grows out of the weaving together of many influences, factors and conditions, all drawn from a broader systemic context. Strangely enough, the same can be said of the mediator's story which, in spite of truly Herculean efforts to maintain neutrality, is brought to the table and enters into the efforts to move beyond the conflict.

The idea that mediators can separate the process issues from the substantive issues is not supported by reality. In some way, this dilemma can be regarded as the mediator's dilemma – a real-life application of Heisenberg's uncertainty principle in quantum mechanics (the inability to know with precision, at a given point in time, both the position and the momentum of a subatomic particle). A mediator cannot focus on process questions uniquely without also influencing and potentially modifying substantive issues that are central to a given dispute.

The issues discussed are only three of the many philosophic challenges to an interest-based approach (for more detail, see Mayer, 2004; Winslade & Monk, 2000). It is beyond the scope of this chapter to explore the many others. Suffice to say that incorporating a relational or narrative approach to addressing conflict in a CAS corresponds more closely to the ways in which such systems develop and evolve. This approach is an even better match in health care CASs, as will be seen later.

Untangling Conflict in Health Care

We have arrived at the point where our senior health care leader is getting a bit impatient, yearning for concrete advice to help her apply the general principles provided by the oracle at Delphi to the dysfunctional silos and mini-kingdoms in her health care system. Fortunately, help is at hand in the form of a series of six articles published in *The American Journal of Nursing* (Gerardi, 2015a–f). These outline in detail how to apply a relational or narrative approach to conflict in a CAS. Gerardi raises two fundamental points with respect to applying such an approach in health care disputes.

First, providing care is a relational activity (Letiche, 2008), depending on the ability of all providers to work collaboratively by building strong relationships within and between teams and departments in order to connect with the needs of the patient, client, family and community of which they are a part. This applies equally to providing care, developing and expressing empathy, organising appropriate support and the like for a specific patient and/or family. The central relational nature of caring is true whether this is provided in a simple therapeutic dyad or a more complicated structure; in other words, health care is fundamentally a relational enterprise even when it is not provided in a CAS setting. Thus, it makes sense that a perspective that sees conflict as something to be broken down into parts that can be fixed or problems to be solved is unlikely to be successful in situations that depend on the development and strengthening of relationships at all levels.

Second, Gerardi (2015a–f) and Mayer (2004) argue persuasively that focusing on the resolution of conflict misses entirely the need for individuals, teams and systems to be actively engaged in addressing the patterns of behaviour that reinforce and sustain conflict in a CAS. This leads to the important concept of conflict engagement, which allows participants to be better prepared to address conflict and disputes arising in the workplace. Engagement will entail understanding and reflecting on patterns of behaviour that influence individuals, teams and systems in their responses to conflict; all of this with a view to becoming more aware of their contributions as they interact dynamically with other agents and components in the system, at multiple levels. Conflict engagement methods and approaches will inevitably lead to different interactions and ultimately to the emergence of new patterns of behaviour.

Unfortunately, many organisations have not yet understood the essential nature of conflict engagement and have not created the possibility for this to develop and be sustained at the individual and systemic level. This undoubtedly reflects the overwhelming tendency to think and react in a linear manner, to see conflict as a 'problem to be solved', and to imagine that the solutions will come from 'fixing the components' whose internal needs and interests have produced the situation in the first place. This perspective is radically different from one that understands conflict as originating in the very nature of the system itself.

Implicit in the application of a conflict engagement approach coupled with relational and narrative mediation efforts is the need for a sustained and effective education and training effort at the organisational and systemic level. This concept has been discussed in detail by Costantino and Merchant (1994) as part of their work in the field of conflict management system design. They emphasise the need for both broad-based systemic training (awareness education) as well as targeted more intense training for identified front-line operators and safety champions. Whether using an Interest Based Negotiation (IBN) approach (see Chapter 3 by Johnson et al., for another example of the value of broad-based educational efforts in facilitating resolution of a conflict) or a narrative relational approach, it is important to address the need for system-level education. Costantino and Merchant's (1994) book is a gem in this regard.

The series of six articles by Gerardi (2015a–f) stand alone in providing useful, concrete advice about approaching conflict within health care CASs. I encourage readers to review the articles in detail (and sequentially). While these have been written to promote understanding and skills within the nursing community, they are immediately relevant not only to the full spectrum of caregivers in a health care setting but also to managers and leaders.

Conclusion

In consideration of these lessons, it is possible to de-construct the oracle's advice as follows:

- Conflict is a natural characteristic of a CAS.
- Health care is a typical, albeit somewhat more complex, example of a CAS.
- There are several distinct ways to approach conflict, some of which are better adapted to CASs and, especially, health care.
- Whether seeking specialised guidance in a particular dispute or developing internal plans to improve an organisation's conflict 'preparedness', select an approach that is suited to your system.
- For most CASs, this will likely involve adopting a relational/ narrative approach to addressing conflict.
- It will always be appropriate to promote conflict engagement at all levels within your system.

References

Bowling, D. & Hoffman, D. (Eds.). (2003). *Bringing Peace into the Room*. San Francisco, CA: John Wiley & Sons.

Braithwaite, J., Churruca, K., Ellis, L. A., Long, J. C., Clay-Williams, R., Damen, N., Herkes, J., Pomare, C., & Ludlow, K. (2017). *Complexity Science in Healthcare: A White Paper: Aspirations, Approaches, Applications and Accomplishments*. Sydney, Australia: Australian Institute of Health Innovation, Macquarie University.

Bush, R. & Folger, J. (1994). *The Promise of Mediation*, San Francisco, CA: Jossey-Bass.

Capra, F. & Luisi, P. L. (2014). *The Systems View of Life*. Cambridge, UK: Cambridge University Press.

Costantino, C. & Merchant, C. (1994). *Designing Conflict Management Systems*. San Francisco, CA: Jossey-Bass.

Fisher, R., Ury, W., & Patton, B. (1983). *Getting to Yes*, 2nd ed. New York, NY: Penguin Books.

Gerardi, D. (2015a). Conflict Engagement: A new Model for Nurses. *The American Journal of Nursing, 115*(3), 56–61.

Gerardi, D. (2015b). Conflict Engagement: Workplace Dynamics. *The American Journal of Nursing, 115*(4), 62–65.

Gerardi, D. (2015c). Conflict Engagement: Collaborative Processes. *The American Journal of Nursing, 115*(5), 66–69.

Gerardi, D. (2015d). Conflict Engagement: A Relational Approach. *The American Journal of Nursing, 115*(7), 56–60.

Gerardi, D. (2015e). Conflict Engagement: Emotional and Social Intelligence. *The American Journal of Nursing, 115*(8), 60–65.
Gerardi, D. (2015f). Conflict Engagement: Creating Connection and Cultivating Curiosity. *The American Journal of Nursing, 115*(9), 60–65.
Hollnagel, E. (2015). Why is Work-as-Imagined Different from Work-as-Done? In R. Wears, E. Hollnagel, & J. Braithwaite (Eds.), *Resilient Health Care, Volume 2* (pp. 249–264). Farnham, UK: Ashgate Publishing.
Kritek, P. (2002). *Negotiating at an Uneven Table*. San Francisco, CA: John Wiley & Sons.
Letiche, H. (2008). *Making Healthcare Care*. Charlotte, NC: Information Age Publishing.
Marcus, L. J., Dorn, B. C., & McNully, E. J. (Eds.). (1995). *Renegotiating Health Care: Resolving Conflict to Build Collaboration*, 2nd ed. San Francisco, CA: Jossey-Bass.
Mayer, B. (2000). *The Dynamics of Conflict Resolution*. San Francisco, CA: Jossey-Bass.
Mayer, B. (2004). *Beyond Neutrality*. San Francisco, CA: John Wiley & Sons.
Moore, C. (1996). *The Mediation Process*, 2nd ed. San Francisco, CA: Jossey-Bass.
Olson, E. E. & Eoyang, G. H. (2001). *Facilitating Organization Change*. San Francisco, CA: Jossey-Bass/Pfeiffer.
Perrow, C. (1984). *Normal Accidents: Living with High-Risk Technologies*. Princeton, NJ: Princeton University Press.
Robson, R. (2015). ECW in Complex Adaptive Systems. In R. Wears, E. Hollnagel, & J. Braithwaite (Eds.), *Resilient Health Care, Volume 2* (pp. 177–188). Farnham, UK: Ashgate Publishing.
Stone, D., Patton, B., & Heen, S. (1999). *Difficult Conversations*. New York, NY: Penguin Putnam.
Ury, W. (1999). *Getting to Peace: Transforming Conflict at Home, at Work, and in the World*. New York, NY: Viking Adult.
Winslade, J. & Monk, G. (2000). *Narrative Mediation*. San Francisco, CA: John Wiley & Sons.
Winslade, J. & Monk, G. (2008). *Practicing Narrative Mediation*. San Francisco, CA: John Wiley & Sons.

Part III

Theorising About Boundaries

5

'Practical' Resilience: Misapplication of Theory?

Sam Sheps
University of British Columbia

Robert L. Wears
University of Florida
Imperial College London

CONTENTS

Introduction

Science has always been challenged by the question of whether investment in theory development and its application will yield practical benefits. While this challenge is perhaps less acute in theoretical physics or pure mathematics, it pervades virtually every branch of health care research. The dramatic emergence of knowledge synthesis, translation and transfer as a fundamentally important research area illustrates the need for a practical application and return on investment. This challenge is crucial to understanding the potential of resilience thinking in the context of patient safety.

The call for the practical application of resilience theory in health care (Feeley, 2017) is relatively recent, in part because the theory itself remains incomplete. However, despite its emergence from an engineering context – which itself required the practical application of a number of scientific domains involving macro- to nano-technologies – addressing the complexities inherent in the institutional context of health care presents a very different challenge. In health care – where clinical science (being dominant in the patient safety movement), jostles with sociology, psychology, political science, communication and organisational perspectives for explanatory power – significant questions regarding what is meant by the *practical application* of resilience arise because of the fundamental nature of resilience as, conceptually, a potential phenomenon. While the notion of practical application in health care is clearly related to practices of many kinds, a question remains: at what level of these practices is resilience particularly relevant? The overall objective of this chapter is to explore this problem. While the history of the patient safety movement over the last 15 years has led us to be concerned about the likely misapplication of resilience as merely another tool in the toolbox, we claim it has much more to offer than that.

Traditionally resilience, in materials science for example, has been conceptualised as a *measurable property* of a specific object – a bar of iron, say. However, in individual, group and organisational contexts, it is clear that resilience is not a measurable property. Thus, at this stage in resilience theory development in health care, it is just not possible to say, for example, that one can increase resilience by 10%. Moreover, resilience has been theorised, in the social and organisational context of health care, as an emergent phenomenon. Thus, because its effects can only be seen subsequent to its emergence, one could posit that resilience doesn't logically exist until then. Hence, resilience as a theory was largely, and significantly, used as a descriptive lens through which to understand the nature of failure, and more recently success, within complex adaptive systems. It was also used to identify elements of the dynamic contexts and behaviours that might have prevented or mitigated failure, or enhanced successful practice. This has been important to understanding those exemplars that can be interpreted as resilient.

In our view, resilience might be better characterised as an emergent *ethos* of Work-as-Done. It is not explicit and is not observable or measurable, per se. Analogous to lightning as an observable event, the underlying cause (electrostatic energy) is not seen, but the potential for electrical discharge exists. Consequently, resilience thinking is shifting towards understanding capacities rather than cataloguing specific discrete actions (Bergström, van Winsen, & Henriqson, 2015). A system 'poised to adapt' (Cook, 2016), and capable of doing so, might be considered potentially resilient. (Note that we do not say such a system *possesses* resilience, as resilience is not an attribute or possession, but rather a form of dynamic emergence).

Critically, resilience in health care and other complex work settings emerges from interactions among and between people and technological and social artifacts, as well as the organisational environment. All such interactions are challenged by boundaries: personal, professional, technical, hierarchical, political and more. The keys to successful practice, given these boundaries (and the often unpredictable interactions across them), are:

- Communication (verbal and non-verbal)
- Environmental scanning for threats and opportunities
- Observing the distribution and flow of tasks comprising the Work-as-Done
- Sensing how the work is proceeding
- Making relevant adjustments as needed to manage existing challenges relating both to the tasks at hand and the conditions in which they occur.

In resilience theory, aspects of these processes are known as the four capabilities: to understand the *actual*, anticipate the *potential*, *monitor* continuously and *learn*. These capabilities have been found to be associated with the emergence of resilience. Yet the instrumental application of resilience in health care has been historically problematic. Our chapter will outline approaches to resilience thinking that address the barriers that hamper efforts to understand and ameliorate patient harm.

Background

Attempts to understand how failures occur, how they might be prevented or mitigated and to explain the suffering that results have preoccupied human kind for millennia (Dekker, 2015). We have gone from blaming failure on angry Gods, to an angry God, to technical, human and organisational weaknesses of many kinds in every domain of human activity. The response to failure has entailed the development of specific acts of supplication, contrition or mitigation: human or animal sacrifice, prayer and payment for dispensation were once deemed practical solutions. (Dekker, Long, & Wybo, 2016). In the pre-Enlightenment period, the notion of the 'fatal flaw' (initially conceptualised as sin, and then less morally laden, as hubris) provided another intriguing explanation for human failure. These explanations tended to dismiss ways of preventing failure other than by 'right living'. Subsequently, Enlightenment rationality posited an orderly universe in which cause and effect were discoverable by sufficiently vigorous application of knowledge and logic; thus all accidents should eventually be explicable, and if explicable, then preventable. Failures were re-conceptualised as being due to an insufficiency of scientific knowledge, the practical solutions for which were research and education. In the technical age, initial solutions to failure were sought in improved design and a rational approach to the organisation of workers (Wears & Hunte, 2014), and ultimately to considerable rhetoric about 'systems'. This has led to organisational 'solutions' that are highly scripted and bureaucratic, which in and of themselves have created challenging barriers to any broader consideration of how failure and success arise (Dekker, 2014). Thus, rationalist ideas about failure remain fundamentally inadequate, their goals ever-receding. Such conceptualisations inevitably become a profound source of organisational anxiety and a liability challenge, which results in even more rigidity and bureaucracy (Beck, 1992).

In the health domain (both clinical and organisational), aside from the ancient injunction to 'do no harm', the issue of patient harm was seen (from an institutional and professional perspective) as normative, an unfortunate but irreducible side effect of advances in treatment of disease and hospital care (Barr, 1955; Mills, 1978; Mills, Boyden, & Rubamen, 1977; Moser, 1956; Ogilvie & Ruedy, 1967a, 1967b; Schimmel, 1964). However, a profound discursive shift was slowly taking place. In 1991, one of the four results papers stemming from the Harvard Medical Practice Study (Leape et al. 1991)[1] coined the term 'error' to describe medical injury, and in 1999, these ideas burst onto the public policy stage with the publication of *'To Err is Human'* by the Institute of Medicine (Kohn, Corrigan, & Donaldson, 1999).

The re-expression of what had long been known as a 'harm' problem to an 'error' problem was a stunning surprise, and met with the astonishment, dismay and calls for action typical of moral panics (Cohen, 1972). This focus on human volition as the cause of harm (a view that other complex industries were beginning to abandon, or at least concede was one of many causes) says much about the superficial state of thinking about safety in health care. As is often the case in dealing with failure, the immediate reaction at higher administrative levels was shock, followed by anger, then fear (usually of liability), accompanied by a serious loss of confidence in the organisation itself and a hunt for 'those responsible'. The first practical action, aside from disciplining or firing the culprits, entailed the establishment of bureaucracies to study and manage the problem: hence the formation of the Canadian Patient Safety Institute, the Agency for Healthcare Research & Quality's Center for Patient Safety in the United States and the UK National Patient Safety Agency (now defunct), which could be seen to be doing something about the problem. The effectiveness of these moves is unknown (and will likely never be evaluated in any real sense), but the political/liability imperatives have been served.

The health professions' response has largely, and not surprisingly, followed a diagnosis and treat process – a very physician-centric approach. This has led to innumerable highly specific activities more or less resembling

[1] Interestingly, the methods paper for the Harvard Medical Practice Study published 2 years earlier included no plans to use 'error' as an outcome, but was based on legal liability (Hiatt et al., 1989); the decision to change the outcome measure during the course of the study has never been explained.

'whack-a-mole'. Other conceptualisations of iatrogenic harm, the Swiss Cheese Model for example, have been somewhat helpful in at least broadening the consideration of cause (from the narrow but highly popular view of root cause) to include human factors related to the use of complex technology, policy and procedure conundrums and trade-offs generated at senior administrative levels. But health care remains trapped in linear, chain-of-events causal thinking. More modern views of accidents (and good performance) as resulting from highly dynamic and unpredictable interactions in complex adaptive systems were not considered in the instrumental pursuit of human failure. Similarly, uninformed rhetoric about systems created an intellectual vacuum, which in turn led to problems in understanding issues of accountability, responsibility and authority.

Thus, the fundamental conceptual basis for the patient safety movement became the measurement and elimination of failure (often cast as 'never events'), which ironically occurs most of the time. The entire patient safety movement was thus framed as a 'practical' approach to the problem of harm, underpinned by vague notions of a reporting, learning and safety culture[2] to grapple with problems of failure that had been long known and addressed with increasing sophistication, in particular, multi-disciplinary engagement in other complex, hazardous industries. However, it had a hard time acknowledging that this wider experience had any real relevance or application to patient harm. These limitations in thinking and practice have trapped health care on a hamster-wheel of patient safety activity (Morrison & Smith, 2000).

Given that theory proposes resilience as emergent, it is immediately clear that any attempt to make resilience instrumental or 'practical' is problematic on a fundamental level. Resilience simply doesn't exist as an entity to which concrete instrumental action can be applied. As noted earlier, it exists rather as a deeply embedded potential. Thus, to think about 'practical resilience', to use the normative idea of practical, is an oxymoron. An emphasis on quick fixes and best practices, policy/procedures development and/or education and credentials can only logically be applied to Work-as-Imagined. It is an approach that codifies specific actions into sets of new rules in the vain attempt to reduce harm through certification and competency efforts, tool kits, road maps and endless highly focused pilot projects which, while superficially satisfying, merely engender a broader set of fantasies about professional practice. Practitioners and administrators will fall into the trap of thinking they now have solutions and will cease to think (or care) more deeply about how work is actually done. Moreover, since one can never know when the operating point is near the safety margin until it has been crossed

[2] Reporting, in particular, is seen not only as the basis of learning (dubious at best given significant under-reporting, virtually non-existent analysis and poor administrative feedback) but also as a way to create targets for improvement. As Goodhart noted: 'When a measure becomes a target, it ceases to be a good measure' (Goodhart, 1975).

(and indeed this may not be immediately clear even after crossing it), it is a logical impossibility to specify *a priori* concrete actions that will unambiguously prevent crossing it, except of course to stop doing the work at all.

Building the Case

Given the preceding characterization of the safety movement generally and of resilience specifically, it is instructive to review several general themes that support the claim that practical resilience is profoundly problematic.

The Fundamental Nature of Organisations

A critical aspect of organisational behaviour that both challenges the practical application of resilience and ironically creates an urgency for a much deeper engagement with the concept is the common view of organisations: clear divisions of labour and responsibility, well-defined fixed and specific communication paths, activities bounded by procedures, hierarchical authority and data driven. In such a linear view, organisational operations and change are perceived as orderly and technical in nature and therefore amenable to being fixed when problems arise (Graetz & Smith, 2010; Weber, 2015). However, this view also highlights the myriad of boundaries that frustrate easy solutions. Thus resilience, as described earlier, may be a relevant alternative way of thinking about how to manage these boundaries that are often much more ambiguous and fluid than current organisational descriptions acknowledge.

The Nature of Practice

In a paper discussing her engagement with Schön's theory of reflective practice, Kinsella (2007) represents a similar challenge to conventional thinking. A working physiotherapist, Kinsella, notes that Schön's theory – at heart a critique of technical rationality – reflects her own experience of 'what it is really like' in practice as well as the complexity of that experience. She notes that Schön (1987) posits technical rationality as an 'epistemology of practice derived from positivist philosophy' (p. 3). She draws on Wilson and Hayes (2000), noting

> Schön's analysis of the 'crisis in the professions' remains the single most poignant depiction of the deep crevices between how we think professionals carry out their work and what working conditions are actually like.
>
> **(p. 104)**

The emphasis here is on the dynamic nature of work as well as the issue of Work-as-Imagined vs. Work-as-Done. The emphasis in Schön's critique regarding the instrumental nature of professional work provides a basis for being wary of normative characterisations of such work as essentially applications of scientific truth (Schön, 1982). Moreover, Schön (1987) also noted that

> In the varied typography of professional practice, there is a high hard ground overlooking a swamp. On the high ground, manageable problems lend themselves to solutions through the application of research-based theory and technique. In the swampy lowland, messy, confusing problems defy technical solutions.
>
> **(p. 103)**

Kinsella (2007) observed that given this 'messy' aspect of practice as 'uncertain', 'unstable', and prone to 'value conflicts',

> … practitioners bound by a positivist epistemology find themselves caught in a dilemma. Their definition of rigorous professional knowledge excludes phenomena central to their practice…the model of technical rationality fails to account for practical competence in divergent situations.
>
> **(p. 106)**

It is just such 'practical competence' (or expertise) that is central to resilience thinking. Clearly, Kinsella is not advocating a quick fix here, but rather describing a need to develop deep-seated potential capacities that can be called upon to respond to uncertainty or surprise.

The Problem with Problem-Solving

Problem-solving ability is, generally, thought to be underpinned by the *successful application of practical solutions*, the accumulation of which is critical for individual and organisational learning and thus successful work. This seems superficially reasonable and, indeed, 'organisational learning' has become a cornerstone objective (primarily rhetorical) of the patient safety movement. However, as Tucker and Edmondson (2002) highlight, problem-solving can have significant negative effects on learning. Their field study of hospital nurses identified that problem-solving was, in effect, *'[reinforcing] the tendency to engage in short-term fixes and thereby limit organizational learning from front-line failures'*.

In reviewing the literature on organisational problem-solving, Tucker and Edmondson (2002) describe the useful distinction, citing a seminal paper by Argyris and Schön (1978), that can be made between first-order and second-order solutions: the former emphasise quick fixes, but with little understanding of the causal dynamics, while the latter involve deep investigation into those dynamics to prevent recurrence. Conceptually, first-order problem-solving is a type of 'workaround'[3], and according to the authors, is often considered a highly prized example of professional competency. This can lead to what Levitt and March (1988) termed the 'organisational competency trap' in which

> ... organizations fail to learn because of a flawed tacit assumption that current competencies are preferable to alternatives.

Of course, the issue of competencies is heavily laden with metrics of dubious value that also inhibit a deeper understanding of Work-as-Done. Moreover, a rigid view of competencies creates boundary issues both within and across professional groups. These boundaries ultimately hamper both communication and the opportunity to make sense of latent or unfolding dynamics in complex systems that might mitigate or prevent patient harm.

The Orthodox Safety Movement Approach

In 2008, a wide-ranging paper 'The epistemology of patient safety research' appeared in the *International Journal of Evidence Based Healthcare* (Runciman et al., 2008). In fact, the paper discusses in detail the epistemology of what Hollnagel (2014) charmingly calls 'accidentology', or what goes wrong. While it is noted that a number of conceptual models regarding patient safety exist, there is no mention of resilience or high reliability. A long list of data sources is noted (Table 2, p. 480; Runciman et al., 2008), and the core methods suggested highlight root cause analysis and plan-do-study-act methodologies. Table 4 (p. 483; Runciman et al., 2008) provides a list of organisations giving priority to patient safety. Each of these is heavily invested in, and by-and-large accepts as valid, standard positivistic approaches and the search for 'safety' metrics to address the problem of patient harm; no outside-the-box thinking here. The authors note they 'support the pragmatic approach' advocated by Leape, Berwick, and Bates (2002):

> Policymakers must consider the entire experience with safety practices both in healthcare and other industries, when deciding which practices should be recommended for widespread use. Evidence from randomized controlled trials is important information, but it is neither sufficient nor necessary for acceptance into practice. The prudent alternative is to make reasonable judgments based on the best available evidence combined with successful experiences in health care.
>
> **(p. 107)**[4]

[3] Workarounds are regarded as an important and successful means of getting necessary work done in complex adaptive systems where surprise, uncertainty, trade-offs and heavy workloads are common. This is one of the few descriptions of workarounds that raise important questions about whether they should be seen *only* in a positive light. See Wears and Vincent (2013) or Cook (2013) for additional thoughts on the 'dark side' of resilience. That said, it is interesting that workarounds have been observed when things go right as well as wrong (Dekker, 2018).

[4] We are not convinced that Leape is referring to fostering successful work, that is Safety-II, but to a successful reduction in error (Safety-I).

Thus, the authors support the continuation of the approach to patient safety that has been less than overwhelmingly successful (see later) and fails to seriously consider a re-conceptualisation of the approach to the patient safety problem. The authors acknowledge some 'fundamental challenges' including 'how to tie down the elusive concept of patient safety' itself, as well as definitional problems of error, violations, system failures, near misses, though interestingly not the consistently vague notion of 'safety culture', which they invoke as being of great importance. Moreover, the authors assert

> A pragmatic approach is needed that recognizes what is obvious and matches the tools of data collection and analysis to the questions that need answering ... The scope of what could be addressed is huge; nevertheless, lists have been developed by influential organizations [see above]... Some of the problems can be 'designed out' quite simply... However using a mix of quantitative and qualitative methods, using information from all available data sources and combining retrospective, real-time and prospective study designs **may** [emphasis added] be necessary to address some of the more difficult patient safety problems.
>
> **(p. 476)**

The paper validates the positivist diagnose and fix perspective that is strongly entrenched in the clinical world and fundamentally assumes that absolute control over highly complex and dynamic processes of patient care and organisational interactions is possible. It fails to acknowledge the importance of boundaries, and thus does not consider that practical approaches attempting to provide that control are insufficient.

Reflections

The differing themes discussed may lead us to think that there are multiple fundamental reasons why an emphasis on the 'practical' application of resilience is not the way forward. And, in our view, they also engender a form of resistance to the very essence of resilience as latent and emergent; to be stimulated by challenges and surprises and capable of enhancing adaptation to them.

First, organisations, particularly health care organisations, are not predisposed to spend time on things that can't be seen – though this may also reflect what they look for. From a senior administrative perspective, solutions to safety problems have to be empirically proven: quick fixes, discipline and avoidance of liability through rules, metrics, procedures and policy are seen as paramount. Trying to understand the deeper, more subtle aspects of how organisational work is actually done is not a high priority. Thus, it is not surprising that resilience is met with puzzlement and is generally ignored as a conceptual basis for thinking about safety. Indeed, even Safety II is generally seen as a result of the implementation of rules and procedures, and rigid adherence to guidelines and best practices; deeper inquiry regarding how things go right is therefore generally considered unnecessary.

Second, as discussion about the unintended consequences of problem-solving on the front-line reveals both the drive for efficiency and pride in professional know-how, act to reinforce quick solutions. There is no time for reflection, communication, mutual support or sense making when problems arise. Practical solutions are honoured only, and especially, if they do not take any time to implement. There is a failure to learn and prevent.

Third, the field of health care continues to demonstrate a remarkable lack of curiosity as to why there has been only modest progress since the Institute of Medicine (IOM) report. For example, in contrast to the Runciman et al. (2008) paper previously discussed, four independent organisations produced separate reports in 2015, all concluding that the patient safety effort since 2000 had not accomplished much and that new thinking about safety was desperately needed (Baker & Black, 2015; Illingworth, 2015; National Patient Safety Foundation, 2015; Pronovost, Ravitz, Stoll, & Kennedy, 2015). Although these four reports accurately characterised the nature of patient safety's failure to make progress, they failed to propose any effective solutions, falling back on the same pleasant sounding but empty rhetorical flourishes that appeared in the IOM report: a 'systems' approach, and a 'culture of safety'. Thus, the four critiques are internally contradictory, saying that what we've done has not worked, so we should do more of what we said we'd do back in 2000. As Albert Einstein may or may not have observed: 'insanity may be defined as doing the same thing over and over again and expecting different results'.

Finally, to discuss (let alone apply in any practical sense) an idea suggesting that successful operations are a function of something un-measureable, latent and emergent is anathema to technical rationality: it is inconceivable.

What Is to Be Done?

If, as we have suggested, practical resilience is an oxymoron, then what is 'the way forward'? First, it is useful to recall that the concept of resilience, as noted earlier, has been largely descriptive. We have examined situations in which organisations have confronted significant challenges and were able to maintain operations during, and/or recover after, a shock: the Baltimore and Ohio Railroad Museum incident, in which a heavy snowfall seriously damaged the museum (a 19th century roundhouse) shortly before a planned festival that was ultimately successfully staged (Christianson, Farkas, Sutcliffe, & Weick, 2009) is a good exemplar, as is the way hospitals in Jerusalem managed mass casualties after a bus bombing (Cook & Nemeth, 2006). Comprehensive, nuanced descriptive research needs to continue to provide further understanding of what resilience looks like when it emerges and what behaviours underpin that emergence in differing circumstances.

Second, resilience in the health care professional, as well as in the ethnographic and sociological (i.e., organisational) sense, is a form of coping (with problems and opportunities) at base both psychological, cognitive and/or psychosocial. Its manifestation depends, in large part, on a stimulus (unpredictable) calling for an array of emotional, cognitive and behavioural capacities. The degree and power with which it emerges is neither predictable nor measureable. Thus, it is hard to reconcile the theoretical essence of resilience as an emergent phenomenon with the normative practical conceptualisations of resilience in health care organisations.

Third, from insights we have already gained, we can say with reasonable conviction that there are a number of fundamental cognitive and behavioural attributes that have been found to underpin the emergence of resilience in high-risk, complex adaptive organisations. These include:

- Maintaining a constant awareness regarding hazards and opportunities based on a shared sense of ongoing potential vulnerability and being aware of weak signals
- A rich repertoire of possible responses based on multi-disciplinary expertise
- Good communications within and, critically, across boundaries
- An allowance for (indeed encouragement of) adaptation in the face of change
- Acknowledgement of the need for reflection on how challenges are recognised and prevented or managed

Each attribute requires a sustained commitment to foster these organisational and individual behaviours. As noted in a recent discussion of anticipation, van Stralen (2017) observes that such behaviours

> ... lead[s] to self-organizing responses to the unexpected (itself a self-organizing process). Behavior [is critical] because we do not [cannot immediately] change the situation, context, or environment with ideas or through thinking. We change things by moving forward with engagement and enactment...[this] involves cognitions (beliefs and ideas), affect (value, attitudes, regulated emotion and fear), and context (the particular and the environment surrounding the person and problem).

Fourth, these are not context-tied or time-limited solutions, but general attitudinal and behavioural changes that must continuously be exercised. They must also acknowledge the effect of pervasive and complex organisational dynamics, including boundaries, to enhance the likelihood of managing adversity and promoting success (Flach & Voorhorst, 2016). Action plans, high-profile sloganeering or 'visioning' will not accomplish such change. Indeed, studies of improvement in health care organisations (and we would claim in discussions of safety as well) have suggested that the rhetoric of empowerment and 'bottom-up' improvement is often merely a smoke screen obscuring managerial, top-down initiatives (Waring & Crompton, 2017): such rhetoric exacerbates the boundary problem rather than mitigates it. As Braithwaite, Runciman, and Merry (2009) note

> The emphasis should be on guiding the natural properties and behaviours of sociotechnical systems [and we would note of the individuals working within them]... rather than on imposing hierarchical structures and above-down instructions from people who do not work at the coal-face.
>
> **(p. 39)**

In summary, we argue that the effort to operationalise resilience (to measure it or make it practical) is inconsistent with its essentially emergent nature. We are concerned that the continuing initiatives to reduce failure in health care practice (both administrative and clinical, i.e., Safety I) have been only marginally successful. Resilience theory emphasizing Safety II changes the focus to 'how things go right' and highlights the vigilant and interactive behaviours noted earlier as critical for resilience in practice to emerge (Dekker, 2018). However, it will take significant change in how failure and success in health care are understood as well as a reconsideration of what is meant by practical. It is not, in our view, the urgent normative drive for targeted but superficial solutions or metrics (often misleading), but a deeper engagement by health care organisations at all levels that will foster the behaviours embedded in everyday practices associated with the emergence of resilience. This, in our view, is the hard practical, most effective work that needs to be done. To paraphrase Chesterton (1910), resilience has not been tried and found wanting; it has been found difficult and not tried.

References

Argyris, C. & Schon, D. (1978). *Organizational Learning: A Theory of Action Perspective.* Reading, MA: Addison-Wesley Publishing.

Baker, G. R. & Black, G. (2015). *Beyond the Quick Fix: Strategies for Improving Patient Safety.* Toronto, ON: University of Toronto.

Barr, D. P. (1955). Hazards of Modern Diagnosis and Therapy – The Price We Pay. *JAMA, 159,* 1452–1456.

Beck, U. (1992). *Risk Society: Towards a New Modernity.* London, UK: Sage Publications.

Bergström, J., van Winsen, R., & Henriqson, E. (2015). On the Rationale of Resilience in the Domain of Safety: A Literature Review. *Reliability Engineering & System Safety, 141,* 131–141.

Braithwaite, J., Runciman, W., & Merry, A. (2009). Towards Safer, Better Healthcare: Harnessing the Natural Properties of Complex Sociotechnical Systems. *BMJ Quality and Safety, 18*(1), 37–41.

Chesterton, G. K. (1910). *What's Wrong with the World,* pt 1, ch 5, as cited in Bartlett J., Familiar Quotations, 14th ed. (p. 918). Boston, MA: Little Brown and Co.

Christianson, M. K., Farkas, M. T., Sutcliffe, K. M., & Weick, K. E. (2009). Learning Through Rare Events: Significant Interruptions at the Baltimore & Ohio Railroad Museum. *Organization Science, 20*(5), 846–860.

Cohen, S. (1972). *Folk Devils and Moral Panics.* New York, NY: Routledge.

Cook, R. & Nemeth, C. (2006). Taking Things in One's Stride: Cognitive Features of Two Resilient Performances. In E. Hollnagel, D. D. Woods, & N. Leveson (Eds.), *Resilience Engineering: Concepts and Precepts* (pp. 205–220). Aldershot, UK: Ashgate Publishing.

Cook, R. I. (2013). Resilience, the Second Story, and Progress on Patient Safety. In E. Hollnagel, J. Braithwaite, & R. L. Wears (Eds.), *Resilient Health Care* (pp. 19–26). Farnham, UK: Ashgate Publishing.

Cook, R. I. (2016). *Poised to Deploy: The C-Suite and Adaptive Capacity.* Velocity. Santa Clara, CA: O'Reilly Associates.

Dekker, S. W. A. (2014). The Bureaucratization of Safety. *Safety Science, 70,* 348–357.

Dekker, S. W. A. (2015). The Psychology of Accident Investigation: Epistemological, Preventive, Moral and Existential Meaning-Making. *Theoretical Issues in Ergonomics Science, 16*(3), 202–113.

Dekker, S. (2018, 28 September). *Why do Things go Right?* Safety Differently.com post.

Dekker, S. W. A., Long, R., & Wybo, J. L. (2016). Zero Vision and a Western Salvation Narrative. *Safety Science, 88,* 219–223.

Feeley, D. (2017, 17 February). *Six Resolutions to Reboot Patient Safety.* Retrieved from http://www.ihi.org/communities/blogs/_layouts/15/ihi/community/blog/itemview.aspx?List=7d1126ec-8f63–4a3b-9926-c44ea3036813&ID=365

Flach, J. M. & Voorhorst, F. (2016). *What Matters? Putting Common Sense to Work* (p. 382). Dayton, OH: Wright State University Libraries. Retrieved from http://corescholar.libraries.wright.edu/books/127/

Goodhart, C. A. E. (1975). Problems of Monetary Management: the UK Experience. Vol 1 of Papers in Monetary Economics, Reserve Bank of Australia.

Graetz, F. & Smith, A. C. T. (2010). Managing Organizational Change: A Philosophies of Change Approach. *Journal of Change Management, 20*(2), 135–154.

Hiatt, H. H., Barnes, B. A., Brennan, T. A., Laird, N. M., Lawthers, A. G., Leape, L. L., … William, G. (1989). A Study of Medical Injury and Medical Malpractice. *New England Journal of Medicine, 321*(7), 480–484.

Hollnagel, E. (2014). Is Safety a Subject for Science? *Safety Science, 67,* 21–24. doi: 10.1016/j.ssci.2013.07.025

Illingworth, J. (2015). *Continuous Improvement of Patient Safety: The Case for Change in the NHS.* London, UK: The Health Foundation. Retrieved 12 November 2015, from http://www.health.org.uk/sites/default/files/ContinuousImprovementPatientSafety.pdf

Kinsella, E. A. (2007) Technical Rationality in Schön's Reflective Practice: Dichotomous or Non-Dualistic Epistemological Position. *Nursing Philosophy,* 8(2), 102–113.

Kohn, L. T., Corrigan, J. M., & Donaldson, M. S. (Eds.). (1999). *To Err is Human: Building a Safer Health System.* Washington, DC: National Academy Press.

Leape, L. L., Berwick, D. M., & Bates, D. W. (2002). What Practices Will Most Improve Safety? Evidence-Based Medicine Meets Patient Safety. *JAMA, 288*(4), 501–507.

Leape, L. L., Brennan, T. A., Laird, N., Lawthers, A. G., Localio, A. R., Barnes, B. A., … Hiatt, H. (1991). The Nature of Adverse Events in Hospitalized Patients. Results of the Harvard Medical Practice Study II. *New England Journal of Medicine, 324*(6), 377–384.

Levitt, B. & March, J. (1988). Organizational Learning. *Annual Review of Sociology, 14*, 319–340.
Mills, D. H. (1978). Medical Insurance Feasibility Study. A Technical Summary. *Western Journal of Medicine, 128*(4), 360–365.
Mills, D. H., Boyden, J. S., & Rubamen, D. S. (Eds.). (1977). *Report on the Medical Insurance Study*. San Francisco, CA: Sutter Publications.
Morrison, I. & Smith, R. (2000). Hamster Health Care. *BMJ, 321*(7276), 1541–1542.
Moser, R. H. (1956). Diseases of Medical Progress. *New England Journal of Medicine, 255*(13), 606–614.
National Patient Safety Foundation. (2015). *Free from Harm: Accelerating Patient Safety Improvement Fifteen Years after to Err is Human*. Cambridge, MA: National Patient Safety Foundation. Retrieved 8 December 2015, from http://www.npsf.org/custom_form.asp?id=03806127-74DF-40FB-A5F2–238D8BE6C24C
Ogilvie, R. I. & Ruedy, J. (1967a). Adverse Drug Reactions during Hospitalization. *Canadian Medical Association Journal, 97*(24), 1450–1457.
Ogilvie, R. I. & Ruedy, J. (1967b). Adverse Reactions during Hospitalization. *Canadian Medical Association Journal, 97*(24), 1445–1450.
Pronovost, P. J., Ravitz, A. D., Stoll, R. A., & Kennedy, S. B. (2015). *Transforming Patient Safety: A Sector-Wide Systems Approach: Report of the WISH Patient Safety Forum 2015*. Qatar: World Innovation Summit for Health. Retrieved 18 February 2015, from http://dpnfts5nbrdps.cloudfront.net/app/media/1430
Runciman, W. B., Baker, G. R., Michel, P., Jauregui, I. L., Lilford, R. J., Andermann, A., … Weeks, W. B. (2008). The Epistemology of Patient Safety Research. *International Journal of Evidence-Based Healthcare, 6*(4), 476–486.
Schimmel, E. M. (1964). The Hazards of Hospitalization. *Annals of Internal Medicine, 60*, 100–110.
Schön, D. (1982). *The Reflective Practitioner: How Professionals Think in Action*. New York, NY: Basic Books.
Schön, D. (1987). *Educating the Reflective Practitioner*. New York, NY: Jossey-Bass.
Tucker, A. L. & Edmondson, A. C. (2002). When Problem Solving Prevents Organizational Learning. *Journal of Organizational Change Management, 15*(2), 122–137.
van Stralen, D. (2017, June 14). *High Reliability Organizations (HRO) Conference Call Series: Reliability and Safety as Behaviors* [personal correspondence].
Waring, J. & Crompton, A. (2017). A 'Movement for Improvement'? A Qualitative Study of the Adoption of Social Movement Strategies in the Implementation of a Quality Improvement Campaign. *Sociology of Health & Illness, 39*(7), 1083–1099.
Wears, R. L. & Hunte, G. S. (2014). Seeing Patient Safety 'Like a State'. *Safety Science, 67*, 50–57.
Wears, R. L. & Vincent, C. A. (2013). Relying on Resilience: Too Much of a Good Thing? In E. Hollnagel, J. Braithwaite, & R. L. Wears (Eds.), *Resilient Health Care* (pp. 135–144). Farnham, UK: Ashgate Publishing.
Weber, M. (2015). *Bureaucracy Weber's Rationalism and Modern Society: New Translations on Politics, Bureaucracy, and Social Stratification* (pp. 73–127). London, UK: Palgrave Macmillan.
Wilson, A. L. & Hayes, E. R. (2000). On Thought and Action in Adult and Continuing Education. In A. L. Wilson & E. R. Hayes (Eds.), *Handbook of Adult and Continuing Education* (pp. 15–32). San Francisco, CA: Jossey-Bass.

6

Creating Resilience in Health Care Organisations through Various Forms of Shared Leadership

Lev Zhuravsky
University of Otago
Waitemata District Health Board

Eric Arne Lofquist
BI Norwegian Business School

Jeffrey Braithwaite
Macquarie University

CONTENTS

Introduction

Health care organisations often need to respond to sudden, unanticipated demands for performance and then return to some form of normal operating conditions as quickly as possible, and with minimal performance loss (Cook & Nemeth, 2006). Organisational resilience serves to provide organisations with the ability to meet a wide variety of such demands and rebound from failure, setbacks, conflicts or other threats to well-being that a team may experience (Morel, Amalberti, & Chauvin, 2008). There is, however, always a mismatch between how things are imagined to be done and how they are actually done in practice (Hollnagel, 2014), and this leads to a gap in operational perspective. In reality, the sharp end of operations is invariably complex and continuously changing; a high degree of adaptability is required to succeed, and more importantly, to avoid disaster. But the reality experienced at the sharp end of operations differs significantly from the reality of the blunt end of the organisation, where strategic decisions are made, goals are set and specific procedures for how goals will be achieved are decided. This, too, is complex, but the timescale for activities is longer, and influence is usually more protracted. In addition, there are rarely simple chains of command that extend from the top of the organisation to the sharp end. This means that there are many operational boundaries involved in the delivery of care. This chapter will focus on how and why these boundaries are managed, how they change, and the role of leadership in closing the gaps that develop over time.

Complex Adaptive Systems and the Leadership-Operations Boundary

Health care provides many examples *par excellence* of a complex adaptive system (CAS). CASs can be designed, but only to a certain degree. Such systems cannot be designed in the same sense that a vehicle or industrial process can be specified. This is because CASs have strong tendencies to learn, adapt and self-organise, and things happen on the run. Taken collectively, these characteristics are defined as emergence: a phenomenon whereby structures and behaviours arise from lower level structures and behaviours, and are constantly re-forming. Consequently, the task of managing CASs becomes a challenge because, in effect, the system keeps redesigning itself. In fact, the root constructs of 'leadership' and 'management' have to be viewed differently for CASs than for other types of systems, as it is not a matter of pre-specifying what should be done, with the lower levels simply carrying out prescribed tasks. There are several areas of recent research addressing the issue of leadership in CASs, particularly dealing with vague or unclear operational boundaries. The first focuses on plurality in leadership, which combines the influence of multiple leaders across structural and professional boundaries that are often fuzzy (Denis, Langley, & Sergi, 2012). The second is the idea of relational coordination where reciprocal relationships connect participants across boundaries through caring, timely and knowledgeable responses to emergent situations (Gittell, 2015; Gittell & Douglass, 2012). And finally, there is a relatively new area of research that addresses intergroup leadership that goes beyond traditional leadership theories where leaders closest to the problem take appropriate action (Hogg, Knippenberg, & Rast, 2012).

This is similar to the idea of deference to expertise in High Reliability Organizations (Weick & Sutcliffe, 2001). Each of these areas focus on leadership across boundaries based on relational links, and can be characterised by some form of shared leadership.

In the past 2 decades, scholars have begun to examine the ways in which health care structures and initiatives reflect the properties of a CAS. Braithwaite, Clay-Williams, Nugus, and Plumb (2013), for instance, provide three different examples of specific projects that apply and illustrate underlying complexity principles. The first project investigated CASs at the agent level, exploring mental health professionals' microsystem conceptualisations of patient safety. This 6-month, mixed-methods study of two teams within an Australian metropolitan mental health service concluded that professionals cannot know about the whole system in which they work, but can usually do a good enough job of keeping their part of the system safe using their rule-of-thumb heuristics, and applying bounded, local rationality (Plumb, Travaglia, Nugus, & Braithwaite, 2011). The second project looked at a functioning CAS at the department level and studied an emergency department (ED) as a reverberating, interacting and adapting, meso-level unit. This ethnographic study involved more than a year of observations in the EDs of two tertiary referral hospitals in Sydney, Australia, between 2005 and 2007. The findings showed that emergency clinicians were responsible for guiding the patients' trajectories of care through various overlapping stages that were simultaneously clinical and organisational (Nugus & Braithwaite, 2010). The last study investigated CASs at the system level, deploying a social science research team to the Australian Capital Territory health system, comprising almost 5,000 staff. This study provided a socio-ecosystems perspective on emergent inter-professional practices across heath care, examining the challenges involved in changing health care professionals' attitudes and entrenched ways of working through external means (Braithwaite et al., 2013). These cases highlighted CAS characteristics at varying levels of abstraction, and demonstrated different resilient behaviours, including self-organisation, adaptation, adaptive capacity, herding and networking.

In each of the earlier cases, the common thread was to identify how organisations engineer resilience into their normal routines based on the concepts and precepts of Resilience Engineering (RE) (Hollnagel, Woods, & Leveson, 2006). RE is a fruitful approach to explaining resilience in CASs, including health care organisations. It strives to identify and correctly value behaviours and resources that contribute to a system's ability to respond to the unexpected. As the name implies, the assumption is that resilience can be engineered into CASs to underpin adaptive capacity. As we see it, adaptive capacity is the ability of individuals at the sharp end of operations to step outside scripted rules, regulations and protocols to handle novel situations or surprise. RE recognises that a portion of a system's variability is both unavoidable and potentially beneficial, and posits that it should therefore be leveraged rather than ignored or controlled. The role of management and leadership in promoting resilience in such systems deserves greater attention, and could extend our understanding of organisational resilience.

Resilience and Complexity

The ideas behind organisational resilience have been around for sometime, but their application to health care is relatively recent. The term resilience originates from the Latin word 'resilire', – the ability of an organism 'to jump or leap back' (Fletcher & Sarkar, 2013). In its modern conception, resilience refers to the ability of a system to absorb disturbance and maintain stability (Holling, 1973). Traditionally, resilience comes from two distinct perspectives. Physical science describes resilience as the ability of materials to resume their original equilibrium state after movement or alteration (Lazarus, 1993; Luthar & Cicchetti, 2000). The social–ecological perspective describes resilience as 'the capacity of a system to absorb disturbance and reorganize while undergoing change so as to still retain essentially the same function, structure and identity, and feedbacks' (Walker, Holling, Carpenter, & Kinzig, 2004, p. 1).

More recent approaches to resilience expand the social–ecological definition of resilience to include decision-making processes by engineering in a diversification of capacities so that organisations are prepared to respond to future events (Bernard, 2004; Suddaby, 2010). This requires resilience at several levels of abstraction that include states, traits, processes and outcomes (Fletcher & Sarkar, 2013). Organisational resilience recognises that modern systems are really complex, dynamic-adaptive socio-technical systems experiencing continuous change, and where the new equilibrium state from adaptation is uncertain and variable. Some approaches even challenge the idea that these are systems, and instead, attempt to describe them in order to capture their essential nature, as in 'Tangled Layered Networks' (Woods, 2015). Either way, organisational resilience recognises the perennially adaptive role of individuals and knotted cohorts of people at the sharp end of operations.

The gap between how work is imagined by those who create the structures, processes and directions of operational systems, and how work is actually carried out receives particular attention in the rapidly developing field of RE research. The two terms commonly used to describe this difference are Work-as-Imagined (WAI) and Work-as-Done (WAD). How this gap is interpreted depends on where the perceiver is positioned within the organisation (Lofquist, Dyson, & Trønnes, 2017). Understanding how this gap is created and how it can be managed is important to organisational resilience. RE also recognises that leadership is one of the factors that creates the structural, social and individual factors that sustain organisational resilience.

In many respects, health care system performance can be compared to a wedge that has both sharp and blunt ends. Practitioners work at the sharp end of the system by applying expertise, knowledge and skills to generate results. The blunt end employs organisational leaders, managers, resources and constraints that shape work at the sharp end (Nemeth, Nunnally, O'Connor, Klock, & Cook, 2005). But there can also be a hammer – in the form of external stakeholders (policymakers, funders, taxpayers) – that drives the wedge by placing demands on the delivery of services through the organisational blunt end to the sharp end. While blunt end (management) cognitive work receives much managerial attention, sharp end (operator) cognition is less researched and more difficult to fathom the closer one gets to the pointy end. In a CAS experiencing continuous environmental changes, such as in health organisational settings, WAD on the front-line of patient care is always different from the WAI undertaken by those who manage and lead organisational clusters and units from the executive suite (Hollnagel, 2014).

Gaps and Boundaries

Practitioners adapt their behaviours based partly on their experience and expertise (Braithwaite, Wears, & Hollnagel, 2017; Hollnagel, Braithwaite, & Wears, 2013; Hollnagel, Wears, & Braithwaite, 2015; Wears, Hollnagel, & Braithwaite, 2015) rather than in response to managerial dictates. There are plenty of examples of how WAD by practitioners and front-line support staff is necessarily different from WAI by systems designers and managers (Hollnagel et al., 2013). Opportunities to understand systems arise from examining situations or circumstances in which they fail to be joined up or integrated, by looking at what happens in the system's cracks and gaps. A recent systematic review of research into social spaces, silos and gaps in health care (Braithwaite, 2010) noted the prevalence of the gap phenomenon. Largely, gaps and silos define the social and professional structures that people inhabit in health care. There are many types of gaps, but two predominate: gaps in the hierarchy, a vertical layer of gaps; and gaps in the heterarchy, a horizontal layer of gaps (see Figure 6.1, although both hierarchies and heterarchies are more fluid than are represented here). Heterarchies, especially, are seen by some as a representation of less formal, more flexible manifestations, and can naturally alter according to the perspective of the perceiver.

A key point is that within all health care systems, there are limits to connectivity, and gaps and silos have the potential to drive health care organisations away from collaborative, cross-organisational resilient performance towards the edge of brittleness. Brittleness refers to the edge of system collapse with potential catastrophic surprise. By way of example, preliminary findings from research on resilience following the Canterbury, New Zealand, earthquakes

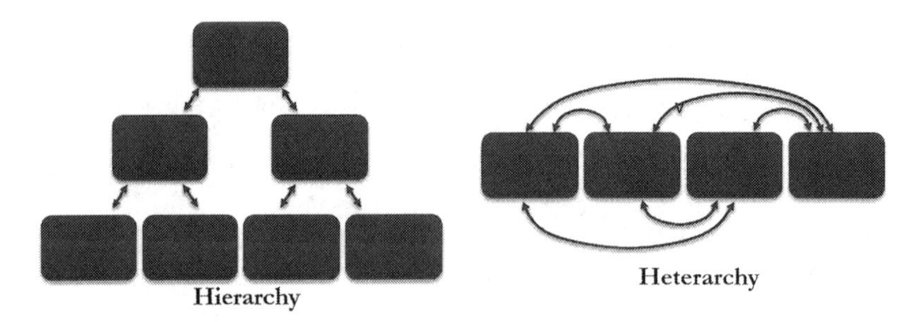

FIGURE 6.1
Hierarchy and heterarchy. (From Braithwaite et al., 2017.)

indicate that adaptive resilience, that is, active responding in the face of crises, requires cross-silo communication and support, and shared leadership (Lee, Vargo, & Seville, 2013). This mirrors growing interest within management and organisational studies for alternative models of leadership in which leadership is not limited to the formally appointed leader but is shaped contingently. Recent studies have also begun to focus on leadership styles that promote resilient behaviour (Lofquist, 2016). In the health care sector, recent attention has been directed towards models of shared leadership – a type of distributed leadership (Greenfield, Braithwaite, Pawsey, Johnson, & Robinson, 2009).

Leading and Managing the Boundaries

Organisational resilience, then, requires individuals and groups that can adapt on the fly, responding not just to disturbances and novelty, but frequent and longitudinal events. This requires a climate that promotes resilient behaviours. Numerous studies have found that some leaders, and specific styles of leadership, can influence the working environment to affect performance (Lofquist, Isaksen, & Dahl, 2017). Yet currently, the role of leadership in creating resilient organisations is little understood. Leaders, by definition, are separated in time and space from the sharp end of operations and often react to a completely different set of challenges.

The aspect of time, in particular, differentiates those at the sharp and blunt ends. External stakeholders, such as governmental authorities, and regulators make decisions and create demands that take place over years. The top leaders then translate these demands into internal processes and procedures that communicate the desired goals of the organisation as well as how they will be achieved, to the sharp end of operations, often through the structure of the organisation and via the distribution of policies, procedures and types of formal prescription. These transmissions can take months or weeks to reach the sharp end, and are often static in nature – e.g., a rigid policy or a standard operating procedure. At the sharp end of operations, on the other hand, where the actual work takes place, individuals must act in a matter of days, hours, minutes, and in some cases, even seconds. This difference in time perspective reinforces the gap that exists between the blunt and sharp ends.

It is not a stretch to form an argument that leadership and management are essentially irrelevant in this process, and that it is economics or professional behaviours, or localised cultural characteristics, that drive decisions. But this view looks at leadership and management as a noun, or a fixed thing – represented as a set of linear interactions and associations within, and across, organisational boundaries, and depicted in organisational charts exercising fixed views and pronouncements. What is less understood are the effects of leadership in an active sense: how leaders' relational interactions affect individual and group performance underpinning resilient behaviours through psychological mechanisms such as those that seek to engage organisational commitment, autonomy, self-efficacy and job engagement (Macey & Schneider, 2008).

Leaders create comfort zones in response to external demands by designing the structure of the organisation, distributing resources, issuing policy, and determining what will be done and, specifically, how articulating goals and achieving them will satisfy external stakeholders. This projection of stability gives them the illusion of control. In a recent interview, the CEO of a large university hospital located in a large city in the United States was asked if doctors always followed the protocols issued? The response was *'absolutely, they must comply with protocols or they will answer to me. I cannot tolerate deviations from the protocols'* (Lofquist, 2018, p. 19). This demonstrated that the leader of the organisation did not really understand the dynamic nature of work at the sharp end. This illusion of control is dispelled at the sharp end of operations: here, individuals do not always respond to upper echelon injunctions, and often act by deviating from, or adapting, prescriptive procedures to achieve satisfactory outcomes. When doctors from this same ED were asked if they always followed protocols, the typical response was *'No way, how could we? Protocols are relatively generic, change regularly, and rarely fit the complexity of the cases we must handle under extreme time pressure and limited resources'* (Lofquist, 2018, p. 20).

Shared Leadership as a Model for Navigating the Gaps

Over the years, leadership research has produced multiple models of organisational leadership, postulating a general assumption: leadership must be exercised by one individual, top-down, via managerial 'prerogatives' to be effective. However, a growing body of research carried out in various sectors, ranging from manufacturing firms to schools to financial sector organisations, has challenged this conventional assumption (Carson, Tesluk, &

Marrone, 2007). Static management models have successfully borne fruit when the leadership task is distributed among team members rather than focused on a single designated leader. This shared leadership is defined as 'a dynamic, interactive influence process among individuals in groups for which the objective is to lead one another to the achievement of group or organizational goals' (Conger & Pearce 2003, p. 286). Traditionally, leadership research has focused on individual leaders and, by extension, on vertical approaches to organising work tasks (Northouse, 2001). Shared approaches to leadership question this individual level perspective, arguing that it focuses excessively on top leaders and says little about informal leadership or larger situational factors. In contrast, shared leadership offers a concept of leadership practice as a group-level phenomenon (Yukl, 2006). Shared leadership can be defined as a dynamic, interactive influence process among individuals in groups for which the objective is to lead one another to the achievement of group or organisational goals (Pearce & Conger, 2003). Although not a novel invention (Gibb, 1954), the theory of shared leadership has only been recently applied in the context of acute health care (Klein, Ziegert, & Xiao, 2006).

The origin of the shared leadership model lies in the research of Benne and Sheats (1948), who suggested that leadership has nothing to do with the individual, but rather with functions, and that several individuals could take up differentiated roles in relation to these functions. It took some time, however, for the model of shared leadership to be applied in the context of acute health care (Flin, O'Connor, & Crichton, 2008). The distribution of leadership in situations with high task load, induced by non-routine events according to skill sets rather than formal leadership ranking in the hierarchy, is very similar to the concept of 'Deference to Expertise' taken from High Reliability Organizations, where decision-making is allocated to the person with the most expertise, and is separate from the formal hierarchy (Künzle et al., 2010). Yet the relationship between shared leadership and team performance cannot be construed in simple terms. Different types of shared leadership relate differently to performance. Mehra, Smith, Dixon, and Robertson (2006) demonstrated that when the team has a distributed–coordinated leadership structure, team performance increases. A distributed–coordinated leadership structure is a form of shared leadership that represent teams that have a formal leader as well as leaders that emerge informally.

Some Examples

In general, then, a shared leadership approach is a dance between two principal components: formal and informal leadership. A network of informal leadership can be a powerful force (Schoenberg, 2004). Informal leadership has been recognised as an important factor in organisational behaviour and team performance (Sink, 1998). There is very little information available that compares interaction of formal and informal leaders in small groups or teams. Most research is conducted on formal leaders or those in a 'position' of leadership (Pielstick, 2000). It is worth noting that, although the importance of shared leadership in low-stress clinical situations is supported by research evidence (Klein et al., 2006), its role in CASs and dynamic clinical environments remains unclear.

The main characteristics of a classic formal leadership role, on the other hand, include decision-making responsibility, influence and authority. However, these leadership forms often evolve into control-based systems that demand rule-based compliance that can stifle or hamper resilient behaviours. Relationship-based approaches, on the other hand, are more likely to facilitate resilient behaviours through psychological–sociological mechanisms, such as encouraging autonomy, self-determination and self-efficacy, leading to extra-role behaviours that promote resilient practices. Relationship-based leadership styles include those described by terms such as transformational leadership, leader-member exchange and authentic leadership. Despite the plethora of descriptors, these often require, and are underpinned by, some form of shared leadership. Shared leadership thus requires leaders to significantly lessen their responsibility for handling local issues by decentralising them to those actually performing the task with an understanding that there will be no reprisals if specific actions fail. This transfer of responsibility creates informal leaders who take personal ownership of their tasks, who have influence and informal authority without necessarily possessing any formal title. Shared leadership, then, is an ongoing fluid process that requires the exercise of continuous support within different contexts, including periods of surprise when novelty or unexpected outcomes arise.

A recent study that explored leadership behaviours in the Intensive Care Unit (ICU) of Christchurch Hospital, New Zealand, following a major earthquake, for instance, suggests that when it came to decision-making, effective communication and the ability to remain calm were the core informal leadership skills and behaviours that led to the successful management and resolution of the crisis (Zhuravsky, 2015). Informal leaders in this study were identified as those individuals who exerted significant influence over other members in their group, although no formal authority had been assigned to them. This research identified four core behaviours attributed to informal leaders: motivation to lead; exercising degrees of autonomy; exhibiting emotional maturity and, understanding that even in a crisis, opportunities present themselves.

This study showed that during the crisis, ICU members adopted a shared leadership approach composed of two main elements. The first involved the sharing of leadership within a formal leadership group that included both medical and nursing sub-groups. The second involved the sharing of leadership between formal and informal leaders across the unit. The conclusion was that if people were to work together in genuinely effective ways, they needed to engage fully in the realities of problem-solving and decision-making in leadership tasks, and be empowered to act with levels of bestowed authority. Most of the participants in this study indicated that sharing leadership tasks helped to create a coordinated, unit-wide response and provided an effective framework for decision-making and overall management of the crisis. It reduced task overload and increased team performance (Zhuravsky, 2015).

Other research has challenged the conventional assumption that leadership must be exercised by one 'heroic' or powerful individual to be effective (Künzle et al., 2010). Shared leadership has thus been advocated elsewhere as an alternative way of improving the performance of teams in other complex environments (St. Pierre, Hofinger, Buerschaper, & Simon, 2011). In this argument, shared leadership, based on building relationships and trust between various levels and organisational ranks, is more likely to promote organisational resilience, creating organisations that are able to rapidly respond to changing circumstances and to improve their performance from those experiences (Stephenson, 2010).

Existing organisational leadership models might be described more precisely as supervisory leadership models (Bass, 1985). Although these models differ in many ways, they presume that the way in which we can explore leadership is to study patterns of employee supervision by single formal leaders (Pearce & Conger, 2003). Seers, Keller, and Wilkerson (2003) suggested that, in contrast to supervisory leadership, shared leadership within teams cannot be based on formal hierarchy and will work better in the environment of a cross-functional team. What distinguishes many cross-functional teams from other organisational forms is the relative absence of formal hierarchical authority. While a cross-functional team may have a formally appointed leader, this individual is more commonly treated as a peer. The purpose of a cross-functional team is to bring a diverse set of functional expertise and experience together. Pearce and Conger (2003) state that cross-functional teams have the advantage of running strong and robust models of shared leadership.

Mielonen (2011) points out that shared leadership is defined as not just a new practical arrangement but a process of working together. This process requires sharing power, authority, knowledge and responsibility. If people are to work together genuinely, they need to engage fully in the realities of problem-solving and decision-making in leadership tasks, and be empowered to act with a certain degree of authority. These attributes should contribute to strong resilient performance across boundaries within organisations and beyond.

Conclusion

Shared leadership approaches have been shown to lead to better outcomes than individual leadership in a variety of contexts, and are considered to be particularly effective in situations of complexity (Knox, 2013). Collective leadership, a form of shared leadership, means a wide range of people take responsibility for the success of the organisation as a whole – not just for their own tasks, jobs or work area. This contrasts with traditional control-based approaches to leadership that have focused on developing individual capability within the context of an organisational setting (West, Eckert, Steward, & Passmore, 2014).

As demonstrated through multiple examples, there is significant potential for supporting resilient health care in complex clinical settings through variations on the theme of shared leadership. Leadership models in this vein can strengthen capacities to traverse professional, structural and cultural gaps, across the multiplicity of organisational boundaries found in health care settings.

References

Bass, B. (1985). *Leadership and Performance Beyond Expectations.* New York, NY: Free Press.

Benne, K. & Sheats, P. (1948). Functional Roles of Group Members. *Journal of Social Issues, 4*(2), 41–49.

Bernard, B. (2004). *Resiliency: What We Have Learned.* San Francisco, CA: WestEd.

Braithwaite, J. (2010). Cultural and Other Associated Enablers, and Barriers, to Adverse Incident Reporting. *Quality and Safety in Health Care, 19,* 229–233.

Braithwaite, J., Churruca, K., Ellis, L. A., Long, J., Clay-Williams, R., Damen, N., … Ludlow, K. (2017). *Complexity Science in Healthcare – Aspirations, Approaches, Applications and Accomplishments: A White Paper.* Sydney, Australia: Australian Institute of Health Innovation, Macquarie University.

Braithwaite, J., Clay-Williams, R., Nugus, P., & Plumb, J. (2013). Health Care as a Complex Adaptive System. In E. Hollnagel, J. Braithwaite, & R. Wears, (Eds.), *Resilient Health Care* (pp. 57–73). Farnham, UK: Ashgate Publishing.

Braithwaite, J., Wears, R. L., & Hollnagel, E. (Eds.). (2017). *Resilient Health Care Volume 3: Reconciling Work-as-Imagined and Work-as-Done.* Abingdon, UK: Taylor & Francis Group.

Braithwaite, J., Westbrook, M., Nugus, P., Greenfield, D., Travaglia, J., Runciman, W., … Westbrook, J. (2013). Continuing Differences Between Health Professions' Attitudes: The Saga of Accomplishing Systems-Wide Professionalism. *International Journal for Quality in Health Care, 25*(1), 8–15.

Carson, J., Tesluk, P., & Marrone, J. (2007). Shared Leadership in Teams: An Investigation of Antecedent Conditions and Performance. *Academy of Management Journal, 50*(5), 1217–1234.

Conger, J. & Pearce, C. (2003). *Shared Leadership.* Thousand Oakes, CA: Sage Publications.

Cook, R. & Nemeth, C. (2006). Taking Things in One's Stride: Cognitive Features of Two Resilient Performances. In E. Hollnagel, D. D. Woods, & N. Leveson, (Eds.), *Resilience Engineering: Concepts and Precepts* (pp. 205–220). Aldershot, UK: Ashgate Publishing.

Denis, J. L., Langley, A., & Sergi, V. (2012). Leadership in the Plural. *Academy of Management Annals, 6*(1), 211–283.

Fletcher, D. & Sarkar, M. (2013). Psychological Resilience: A Review and Critique of Definitions, Concepts and Theory. *European Psychologist, 18*(1), 12–23.

Flin, R., O'Connor, P., & Crichton, M. (2008). *Safety at the Sharp End. A Guide to Non-Technical Skills.* Aldershot, UK: Ashgate Publishing.

Gibb, C. (1954). Leadership. In G. Lindzey, (Ed.), *Handbook of Social Psychology* (pp. 877–920). Reading, MA: Addison Wesley.

Gittell, J. (2015). How Interdependent Parties Build Relational Coordination to Achieve Their Desired Outcomes. *Negotiation Journal, 10,* 387–391.

Gittell, J. H. & Douglass, A. (2012). Relational Bureaucracy: Structuring Reciprocal Relationships into Roles. *Academy of Management Review, 37*(4), 709–733.

Greenfield, D., Braithwaite, J., Pawsey, M., Johnson, B., & Robinson, M. (2009). Distributed Leadership to Mobilise Capacity for Accreditation Research. *Journal of Health Organization and Management, 23*(2), 255–267.

Hogg, M. A., Knippenberg, D. V., & Rast, D. E. (2012). Intergroup Leadership in Organizations: Leading Across Group and Organizational Boundaries. *Academy of Management Review, 37*(2), 232–255.

Holling, C. S. (1973). Resilience and Stability of Ecological Systems. *Annual Review of Ecology and Systematics, 4*(1), 1–23.

Hollnagel, E. (2014). *Safety-I and Safety-II: The Past and Future of Safety Management.* Farnham, UK: Ashgate Publishing.

Hollnagel, E., Braithwaite, J., & Wears, R. L. (Eds.). (2013). *Resilient Health Care.* Farnham, UK: Ashgate Publishing.

Hollnagel, E., Wears, R. L., & Braithwaite, J. (2015). *From Safety-I to Safety-II: A White Paper.* The Resilient Health Care Net: Published simultaneously by the University of Southern Denmark. University of Florida, USA, and Macquarie University, Australia.

Hollnagel, E., Woods, D. D., & Leveson, N. (Eds.). (2006). *Resilience Engineering: Concepts and Precepts.* Aldershot, UK: Ashgate Publishing.

Klein, J., Ziegert, J., & Xiao, Y. (2006). Dynamic Delegation: Shared, Hierarchial and Deindividualized Leadership in Extreme Action Teams. *Administrative Science Quarterly, 51,* 590–621.

Knox Clarke, P. (2013). *Who's in Charge Here? A Literature Review of Approaches to Leadership in Humanitarian Operations.* London, UK: ALNAP.

Künzle, B., Zala-Mezö, E., Wacker, J., Kolbe, M., Spahn, D. R., & Grote, G. (2010). Leadership in Anaesthesia Teams: The Most Effective Leadership is Shared. *Quality and Safety in Health Care, 19*(6), 1–6.

Lazarus, R. S. (1993). From Psychological Stress to Emotions. A History of Changing Outlooks. *Annual Review of Psychology, 44,* 1–21.

Lee, A.V., Vargo, J., & Seville, E. (2013). Developing a Tool to Measure and Compare Organizations' Resilience. *Natural Hazards Review, 14*(1), 29–41.

Lofquist, E. A. (2016). *Resilient Leadership: Exploring the Most Appropriate Leadership Style for Resilient Organizations with in the Health Care Sector.* Resilient Health Care Net Annual Meeting Proceedings. Middelfart, Denmark.

Lofquist, E. A. (2018). *Improving Patient Safety Through Leadership Styles that Promote Resilient Behaviors.* Academy of Management Annual Meeting Working Paper.

Lofquist, E. A., Dyson, P. K., & Trønnes, S. N. (2017). Mind the Gap: A Qualitative Approach to Assessing Why Different Sub-Cultures Within High-Risk Industries Interpret Safety Rule Gaps in Different Ways. *Safety Science, 92I*, 241–256.
Lofquist, E. A., Isaksen, S. G., & Dahl, T. J. (2017). Exploring Change, Job Engagement and Work Environment in the Norwegian Directorate of Fisheries. *Academy of Management Annual Meeting Proceedings, 2017*(1), 12841.
Luthar, S. S. & Cicchetti, D. (2000). The Construct of Resilience: Implications for Interventions and Social Policies. *Development and Psychopathology, 12*, 857–885.
Macey, W. H. & Schneider, B. (2008). The Meaning of Employee Engagement. *Industrial and Organizational Psychology, 1*(1), 3–30.
Mehra, A., Smith, B., Dixon, A., & Robertson, B. (2006). Distributed Leadership in Teams. The Network of Leadership Perceptions and Team Performance. *The Leadership Quarterly, 17*(3), 232–245.
Mielonen, J. (2011). Making Sense of Shared Leadership. A Case Study of Leadership Processes Without Formal Leadership Structure in Team Context (Doctoral Dissertation). Acta Universitatis Lappeenrantaensis, 451.
Morel, G., Amalberti, R., & Chauvin, C. (2008). Articulating the Differences Between Safety and Resilience: The Decision-Making Process of Professional Sea-Fishing Skippers. *Human Factors, 50*, 1–16.
Nemeth, C., Nunnally, M., O'Connor, P., Klock, P., & Cook, R. (2005). Getting to the Point: Developing IT for the Sharp End of Healthcare. *Journal of Biomedical Informatics, 38*, 18–25.
Northouse, P. (2001). *Leadership: Theory and Practice*. Thousand Oakes, CA: Sage Publications.
Nugus, P. & Braithwaite, J. (2010). The Dynamic Interaction of Quality and Efficiency in the Emergency Department: Squaring the Circle? *Social Science & Medicine, 70*(4), 511–517.
Pearce, C. & Conger, J. (2003). *Shared leadership: Reframing Hows and Whys of Leadership*. Thousand Oakes, CA: Sage Publications.
Pielstick, C. (2000). Formal vs. Informal Leading. A Comparative Analysis. *Journal of Leadership and Organizational Studies, 7*(3), 99–114.
Plumb, J., Travaglia, J., Nugus, P., & Braithwaite, J. (2011). Professional Conceptualisation and Accomplishment of Patient Safety in Mental Health Care: An Ethnographic Approach. *BMC Health Services Research, 11*(1), 100.
Schoenberg, A. (2004). *What It Means to Lead During the Crisis: An Explanatory Examination of Crisis Leadership*. New York, NY: Syracuse University Press.
Seers, A., Keller, T., & Wilkerson, J. (2003). Can Team Members Share Leadership? Foundations in Research and Theory. In C. Pearce & J. Conger (Eds.), *Shared Leadership: Reframing the Hows and Whys of Leadership* (pp. 77–102). Thousand Oakes, CA: Sage Publications.
Sink, D. (1998). Who Will Lead the Transformation. *Training, 35*(1), 5–10.
St. Pierre, M., Hofinger, G., Buerschaper, C., & Simon, R. (2011). Crisis Management in Acute Care Settings: Human Factors, *Team Psychology, and Patients Safety in a High Stakes Environment*. London, UK: Springer.
Stephenson, A. (2010). *Benchmarking the Resilience of Organisations* (Doctoral Dissertation). University of Canterbury.
Suddaby, R. (2010). Challenges for Institutional Theory. *Journal of Management Inquiry, 19*, 14–20.

Walker, B., Holling, C. S., Carpenter, S. R., & Kinzig, A. (2004). Resilience, Adaptability and Transformability in Social-Ecological Systems. *Ecology and Society, 9*(2), 5.
Wears, R. L., Hollnagel, E., & Braithwaite, J. (Eds.). (2015). *Resilient Health Care, Volume 2: The Resilience of Everyday Clinical Work*. Farnham, UK: Ashgate Publishing.
Weick, K. & Sutcliffe, K. (2001). *Managing the Unexpected*. San Francisco, CA: Jossey-Bass.
West, M., Eckert, R., Steward, K., & Passmore, B. (2014). *Developing Collective Leadership for Healthcare*. London, UK: The King's Fund.
Woods, D. D. (2015, October). *How Complexity Overwhelms Rules: Building Graceful Extensibility to Manage Surprises. Eurocontrol Human Factors and System Safety Seminar: Understanding Normal Work*. Barcelona, Spain: Eurocontrol.
Yukl, G. (2006). *Leadership in Organizations*. Upper Saddle River, NJ: Pearson Prentice-Hall.
Zhuravsky, L. (2015). Crisis Leadership in an Acute Clinical Setting: Christchurch Hospital, New Zealand ICU Experience Following the February 2011 Earthquake. *Prehospital and Disaster Medicine, 30*(2), 131–136.

7

Simulation: A Tool to Detect and Traverse Boundaries

Mary D. Patterson
University of Florida
Akron Children's Hospital, Simulation Center for Safety and Reliability

Peter Dieckmann
Copenhagen Academy for Medical Education and Simulation (CAMES)

Ellen S. Deutsch
Pennsylvania Patient Safety Authority
The Children's Hospital of Philadelphia

CONTENTS

Introduction

The emergence of the concepts of 'Work-as-Done' and 'Work-as-Imagined' demonstrate that there are borders in which ideal or imagined work is seated and borders within which actual work is accomplished. Depending on the patient care situation, the overlap between these two concepts or the overlap between how people conceive of Work-as-Imagined and Work-as-Done may be substantial or minimal. There is value in understanding that these borders manifest and even more value in creating an understanding of the different ways in which borders may be perceived and work can be accomplished. Managers and administrators who create Work-as-Imagined are often separated by training, role and location from the front-line clinicians who perform Work-as-Done. What is needed is a kind of process to cross the borders between the groups, a 'translator' who speaks the language of both groups or a facilitator who can help participants understand the constraints and pressures that exist for each group.

The nature of simulation and the requisite elements of simulation provide opportunities for health care professionals to reflect on their individual work, their team's work and the work of the system. Both similarities and differences between imagined work, simulated work and actual work trigger such reflections. In the case of similarities, participants might become aware of the actual organisation of work in clinical practice by explicitly discussing it in simulation. In the case of differences, awareness and understanding of how work was accomplished might shed light on the rationale for how work is organised in a clinical setting. This approach might expose and open boundaries of tradition and safety to allow experimentation with new ways of organising work.

Simulations can be conducted in simulation centres or in situ (in the clinical environment) (Lockman, Ambardekar, & Deutsch, 2015), and most commonly include front-line care providers. However, potential participants range from executive leaders to allied health care and support personnel; simulation can help cross professional boundaries. Debriefing, the built-in time for reflection that occurs as a critical element of simulation, facilitates critical thinking (Kihlgren, Spanager, & Dieckmann, 2015) and reflection about health care processes. Importantly, debriefing illuminates the boundaries that exist between different types of professionals (e.g., doctors and nurses) or between clinical units and specialties (e.g., the operating room and the intensive care unit, or the emergency department and the inpatient unit). When those boundaries become noticeable in the debriefing, they can be addressed, managed, minimised, optimised or – at times – removed from the system. In this chapter, we explore three ways simulation can illuminate boundaries and work across barriers. Specifically, we will discuss in situ simulation, debriefing and the simulation of everyday work as a means to understand and work across boundaries.

In Situ Simulation

In situ simulation is likely to be as close as possible to Work-as-Done without being Work-as-Done. In situ simulation involves real teams using real resources in actual patient care environments. Scenarios often reflect typical clinical work, although they may also reflect events that are uncommon but not unimaginable. Participants deal with the presence, absence or ease of access to resources (e.g., equipment, supplies, electronic records), advantages or limitations of the physical environment, capabilities and constraints of the knowledge and technical skills of their team members, and other facets of socio-technical conditions. For example, a simulation may involve a technology-enhanced manikin that develops an arrhythmia and is about to 'suffer' a cardiac arrest. The manikin demonstrates evolving physical findings and vital signs in real time. The clinical team needs to respond and resuscitate the manikin, who, in turn, has different physiologic responses depending on the team's interventions. The participants need to manage with the staff, equipment, clinical protocols and other resources that are – or are not – available in the actual clinical care setting.

Though the 'stem' or initiation of the scenario may be controlled, the health care workers' responses to some sort of perturbation of the patient's condition, or the system, retains the unscripted or 'in the wild' aspect of Work-as-Done. We have previously discussed the ways in which in situ simulation may identify the shortcuts and workarounds that are part of Work-as-Done in any health care system (Deutsch, Fairbanks, & Patterson, 2019). In addition, simulation helps to illuminate professional and discipline-based boundaries that exist between various health care professions or specialties. These boundaries may be seen in examples of inadequate communication between health care workers, absence of common language and presence of authority gradients in a simulation. Simulation also provides insights at the system level. Simulation serves to identify boundary conditions and how these conditions may change as a result of intended or unintended changes in the system.

When evaluating a health care system (for example, a unit-based system, such as a ward or a critical care unit), performance boundaries are often unclear. The boundaries are likely to be dynamic depending on census, staffing, acuity and a multitude of other factors. Though front-line workers may intuitively recognise when they are stretched to and beyond their limits, it may be difficult to define the boundary conditions of performance (Nemeth, Wears, Woods, Hollnagel, & Cook, 2008). More importantly, perhaps, it may not be clear how to define or support those elements that contribute to 'graceful extensibility', which is 'how a system extends performance or brings extra adaptive capacity to bear when surprise events challenge its boundaries' (Woods, 2015).

In situ simulation enables the testing of a system under conditions that very nearly replicate actual clinical care, and yet can be varied systematically to explore system behaviour under different conditions. While in situ simulation is one tool that may be used to explore Work-as-Done, it may be combined with another tool that is often used for a similar purpose: the Functional Resonance Analysis Method (Hollnagel, 2018). FRAM is described as

> A method to analyse how work activities take place either retrospectively or prospectively. This is done by analysing work activities in order to produce a model or representation of how work is done. This model can then be used for specific types of analysis, whether to determine

> how something went wrong, to look for possible bottlenecks or hazards, to check the feasibility of proposed solutions or interventions, or simply to understand how an activity (or a service) takes place.
>
> **(Hollnagel, 2018)**

Combining FRAM and in situ simulation would enable consideration of systematic variation along the six dimensions of FRAM: input, output, resources, preconditions, controls and time. The thoughtful conduct of simulation with experienced facilitators creates an opportunity for the clinical team to understand those elements of the system and those characteristics of the team members that enhance (or hinder) adaptive responses. FRAM provides a model to help explore the relationships that develop or surface during simulations: the aspects can help in designing a range of relevant scenarios as well as the conduct of debriefings.

Surprise is inevitable in health care delivery, and Finkel (2011) has postulated that the ability to recover from surprise depends on what capacities are already present that can be deployed to address the unexpected. In the military, this refers to a diversity of weapons (tools) and responses that mirror biological diversity (Finkel, 2011). Similarly, Finkel (2011) advocates for tolerance and discussion of ideas that challenge the conventional or dominant view. Simulation provides a safe environment in which to 'test' a diversity of responses, including non-standard responses. Testing can then occur without direct risk to patients in a low-stakes setting. This type of system 'stress-testing' has the potential to diminish cultural barriers to non-standard responses, to encourage a flexible and adaptive response to the unexpected. A key component in this approach is to make a clear distinction between the explorative 'idea-testing' phase and actual work.

New processes and technology, especially related to health information technology, are often implemented in health care delivery without understanding the potential effect on the adaptive capacity of the system. In response to an adverse event or undesirable outcome, progressively constraining protocols and processes may be implemented. The result may be an increasingly brittle system and reduction in adaptive capacity. Though workers at the front-line may intuitively understand the change in the system's capacity, it may be less apparent to managers and leadership. The fallacy of centrality, first described by Westrum, describes the belief that if something bad was going on, the supervisor would know since all relevant information flows though the manager (Weick & Sutcliffe, 2007). But there may be many reasons why health care providers don't report, or sometimes even recognise, that a process is problematic. Managers may not be aware or understand the significance of information that percolates to their level. Integrating expert simulation facilitators with the safety and quality infrastructure of a health care organisation, as well as with clinicians, enables facilitators to assist in bridging information gaps.

Allspaw (2012) has described a blameless post-mortem in which problems resulting from 'fixes' to another event are addressed without ascribing blame. In situ simulation has the potential to break down the barriers between administrative and clinical perspectives by explicitly describing the risk and loss of adaptive capacity that occurs as a result of what may seem to be a minor modification to the system. Stresses and limitations that may have been invisible to managers and potentially even front-line workers may be unmasked during simulated conditions. In essence, simulation creates the opportunity to create a blameless pre-mortem, a proactive opportunity to learn.

Though training and simulation specifically have the potential to enhance the repertoire of responses when faced with the unexpected (Clay-Williams & Braithwaite, 2015), the larger contribution relative to the elimination of barriers to optimal responses occurs through increasing understanding of the other – the other person, discipline, team or system. During a recent series of simulated events, interdisciplinary teams of nurses, physicians and respiratory therapists were asked to emergently care for a simulated toddler with a tracheostomy, who was experiencing respiratory distress and cyanosis. Nurses and respiratory therapists in this particular institution are the subject matter experts for tracheostomy emergencies. Yet it was observed that nurses often hesitated to change the tracheostomy without an order from the physician. In contrast, the particular physicians who responded to this emergency had very little knowledge of tracheostomy management. In the debriefing discussion, physicians said that the nurse (with this skill set) should just change the tracheostomy tube rather than wait for an order as the physician did not know what was needed. Nurses expressed surprise that some physicians did not have this particular skill set. Discussion during the debriefing focused not only on tracheostomy management but also on how ad hoc teams might quickly eliminate the boundaries between different disciplines and gain an understanding of the expertise and skills of specific team members.

Debriefing to Break Down Barriers

Simulation provides a space for overcoming boundaries between disciplines, units and specialties. Much of this progress occurs during the debriefing, which is an integral component of learning during simulations. Tannenbaum and Cerasoli (2013) describe the essential elements of debriefing, which include (i) participant engagement in active self-discovery; (ii) a clear, primary intent for improvement or learning that is non-punitive and (iii) reflection on specific events or performance episodes (Tannenbaum & Cerasoli,

2013). Participants work through intense experiences during the simulated scenarios, and thus have a shared experience that may trigger contact points between people who would not typically discuss these matters. It is possible that emphasising psychological safety and conducting the debriefing in a small group setting immediately following the simulation contributes to the effectiveness of this process.

To these essential elements, we consider adding the necessity of expert facilitators. Depending on the goal of the simulation experience, facilitators may represent clinical subject matter experts (nurses and physicians, for example) as well as facilitators with specific expertise in the debriefing process – often the same person brings both aspects of ability to the table. In the example of the simulated patient with a tracheostomy obstruction, the facilitators needed to recognise, expose and explore the boundaries that were uncovered. In this way, the facilitators serve as translators or guides for the diverse clinical team members. Facilitators support the emergence of understanding from the insights of the participants, and the participants develop insights into their own assumptions and capabilities as well as those of others. Explicitly describing and exploring these boundaries during debriefing supports the goal of improving clinicians' performance in the clinical environment. The facilitator as translator is crucial to elucidating the boundaries between groups as well as assisting the health care workers in understanding system performance boundaries.

Areas of emphasis during the debriefing are based on a combination of pre-established goals and learning needs that become evident during the simulation. Often, components of the socio-technical health care delivery system that could be improved are serendipitously identified; insights may arise from both participants and facilitators. At other times, observations about the socio-technical health care system are intentionally sought, such as when simulations are conducted before opening new patient care areas (Deutsch et al., 2016; Geis, Pio, Pendergrass, Moyer, & Patterson, 2011) or when aspects of health care processes are known to be problematic. Intentional emphasis on understanding system processes can be used as a mechanism to deflect attention away from the performance of individuals. Regardless of the focus of the simulation, individual learning is almost inevitable.

The atmosphere of respect that is cultivated during simulation debriefings provides an effective lesson that contributes to breaking down barriers. Different disciplines begin to understand the expertise of the 'other' and that increases their knowledge of the potential for adaptation, the capabilities of team members and the system's built-in resilience and risk. For example, during the debriefing of an emergency department paediatric trauma simulation, the radiology technician said that if the nurse or doctor had just moved over a few inches, the technician would have been able to insert the radiographic plate more easily. The technician may not have had the opportunity or felt empowered to make that request in other circumstances; and it was an easy adjustment for the provider to make. In a similar simulation, it became evident that while the physicians ordered weight-based medications for the patient in milligrams, the nurses measured doses in millilitres; during the debriefing, the participants recognised that a translation process was needed.

This type of debriefing and reflection does not typically occur during everyday work, but it has been occasionally observed. One indication of the value of simulation and debriefing would be to have care providers internalise and incorporate debriefing processes into everyday clinical care, to break down barriers between disciplines, units and specialties – and build adaptive capacity – on a routine basis.

The Value of Simulation of Routine Work

Traditionally, the value of simulation is seen when simulating a certain range of situations such as those that rarely occur; those that are complicated, where many variables and their interdependencies need to be taken into account; complex situations in which the full dynamic is not understood; sensitive situations that do not allow room for much deviation or failure; or situations in which interventions are time critical and for which 'correct' interventions must occur quickly. There are, for example, collections of typical challenges to be simulated for different specialties that build on this logic (Gaba, Fish, Howard, & Burden, 2015). Obviously, this approach makes good use of the possibilities of simulation to practice critical tasks without directly endangering patients, care providers, the health care system or expensive resources. Work by the editors of this book and many others, however, points to the potential to supplement this approach with a conscious focus on everyday work – daily routines and their variability (Dieckmann et al., 2017; Hollnagel, 2014, 2017; Iedema, Mesman, & Carrol, 2013; Wears, Hollnagel, & Braithwaite, 2015).

Supplementing current simulation practice with attention to everyday health care delivery holds several promises. It would address a vast majority of work that is done, as most of this is routine. Because it's rare that events are unambiguously either a major success or a catastrophic failure, we may miss an opportunity to apply learnings in a timely manner, if we wait until the next crisis occurs. The focus on everyday work should also put less cognitive load on the learners (Fraser et al., 2012). There are different types of cognitive load: intrinsic load describes the inherent difficulty of the issues discussed, extrinsic load describes how the material is presented and germane load describes the energy used to process the material to be learned. Our assumption is that routine tasks place a lower intrinsic cognitive load onto the learner, using less of the finite human capacity for processing information, as the issues being managed are common. Decreasing the intrinsic cognitive load may improve the ability of participants to consider how work processes, teamwork, decision-making and so on influence patient care, resource use and other parameters relevant to clinical success.

In addition, it might be psychologically safer (Edmondson, 1999) to simulate and discuss regular practice rather than extreme situations (Dieckmann & Krage, 2013). In addition to the changes described previously, the proposed changes in practice could be expected to improve simulation teams' understanding of everyday practice and its variability. The range of variability in everyday practice and the large 'corridor of normal performance' would become clearer (Dieckmann et al., 2017; Hollnagel, 2014, 2017; Wears et al., 2015). In fact, it would be possible to understand more clearly the boundaries of normal performance. This may make it easier to understand when the boundaries of normal performance are being violated. The conscious insight that a colleague performs the 'same' task just a bit differently will be a tremendous asset when building and using shared mental models (Lowe, Ireland, Ross, & Ker, 2016) that form the basis of co-ordinated care (Bogdanovic, Perry, Guggenheim, & Manser, 2015; Burtscher et al., 2011; Kolbe, Burtscher, & Manser, 2013; Manser, Foster, Flin, & Patey, 2013). Depending on the focus of discussion, such insights could be formulated on the individual, team or organisational level.

The analytical frameworks used in Resilience and Safety-II can be helpful to systematise the design of simulation scenarios and guide debriefing discussions:

- What triggers an action? Which prerequisites are necessary? What guidance is available (and used)? What time aspects are relevant? Which resources are necessary for the process under discussion? What are the outcomes of the process? Are the outcomes timely? Are the outcomes of appropriate quality (Clay-Williams, Hounsgaard, & Hollnagel, 2015)?
- Which of these potential capacities are at risk, well developed or in need of improvement: the potential to react, to monitor, to learn and/or the potential to anticipate (Hollnagel, 2017)?
- How should we balance thoroughness and efficiency? What are the possible trade-offs (Hollnagel, 2009)?
- How do we see the accountability for the different steps of care (Dekker, 2017)?

If these principles are better understood, it will be possible to design simulation scenarios in a way that triggers learning opportunities to build adaptive expertise (Rasmussen et al., 2013). Such a deeper understanding would facilitate a multi-level learning cycle and could result in a simulation practice that focuses on goals rather than rules or algorithms. Especially when conducted across the boundaries of professions and departments, simulation can be the process and translator that can cross borders between groups, speak the language of multiple groups, and facilitate an understanding of the constraints and pressures that shape health care delivery.

Summary

In this chapter, we explored the value of simulation for illuminating and managing boundaries to understand and improve systems from a Safety-II perspective. In situ simulation, in particular, can make system barriers and boundaries visible. Participants understand each other's perspectives better, which may also increase mutual acceptance. On the system level, the simulation team can systematically vary performance conditions and thus understand facilitators and barriers for adaptations. Learning can occur at many levels, ranging from the individual to the organisation – especially when skilfully facilitated by those who consider the different feedback loops needed. Debriefing is the part of simulation practice that allows for reflection and creates the potential for deep understanding. The focus of simulations should extend beyond infrequent and complicated situations, to include everyday uncomplicated work, in order to facilitate learning about and enhancing adaptive practices.

References

Allspaw, J. (2012, May 22). Blameless PostMortems and a Just Culture. Retrieved from https://codeascraft.com/2012/05/22/blameless-postmortems/. Accessed 17 November 2018.

Bogdanovic, J., Perry, J., Guggenheim, M., & Manser, T. (2015). Adaptive Coordination in Surgical Teams: An Interview Study. *BMC Health Services Research, 15,* 128.

Burtscher, M. J., Manser, T., Kolbe, M., Grote, G., Grande, B., Spahn, D. R., & Wacker, J. (2011). Adaptation in Anaesthesia Team Coordination in Response to a Simulated Critical Event and its Relationship to Clinical Performance. *British Journal of Anaesthesia, 106*(6), 801–806.

Clay-Williams, R. & Braithwaite, J. (2015). Realigning Work-as-Imagined and Work-as-Done: Can Training Help? In E. Hollnagel, J. Braithwaite, & R. Wears (Eds.), *Resilient Health Care, Volume 3, Reconciling Work-as-Imagined and Work-as-Done* (pp. 153–162). Boca Raton, FL: CRC Press.

Clay-Williams, R., Hounsgaard, J., & Hollnagel, E. (2015). Where the Rubber Meets the Road: Using FRAM to Align Work-as-Imagined with Work-as-Done when Implementing Clinical Guidelines. *Implementation Science, 10,* 125.

Dekker, S. (2017). *Just Culture: Restoring Trust and Accountability in your Organization*, 3rd ed. Boca Raton, FL: CRC Press.

Deutsch, E. S., Dong, Y., Halamek, L. P., Rosen, M. A., Taekman, J. M., & Rice, J. (2016). Leveraging Health Care Simulation Technology for Human Factors Research: Closing the Gap Between Lab and Bedside. *Human Factors, 58*(7), 1082–1095.

Deutsch, E. S., Fairbanks, R., & Patterson, M. D. (2019). Simulation as a Tool to Study Systems and Enhance Resilience. In E. Hollnagel, R. L. Wears, & J. Braithwaite (Eds.), *Delivering Resilient Health Care* (pp. 56–65). New York, NY: Routledge.

Dieckmann, P. & Krage, R. (2013). Simulation and Psychology: Creating, Recognizing and Using Learning Opportunities. *Current Opinion in Anaesthesiology, 26*(6), 714–720.

Dieckmann, P., Patterson, M. D., Lahlou, S., Mesman, J., Nystrom, P., & Krage, R. (2017). Variation and Adaptation: Learning from Success in Patient Safety-Oriented Simulation Training. *Advances in Simulation, 31*(2), 21.

Edmondson, A. (1999). Psychological Safety and Learning Behavior in Work Teams. *Administrative Science Quarterly, 44*(1), 350–383.

Finkel, M. (2011). *On Flexibility: Recovery from Technological and Doctrinal Surprise on the Battlefield.* Stanford, CA: Stanford University Press.

Fraser, K., Ma, I., Teteris, E., Baxter, H., Wright, B., & McLaughlin, K. (2012). Emotion, Cognitive Load and Learning Outcomes During Simulation Training. *Medical Education, 46*(11), 1055–1062.

Gaba, D. M., Fish, K., Howard, S. K., & Burden, A. (2015). *Crisis Management in Anesthesiology*, 2nd ed. Philadelphia, PA: Saunders.

Geis, G. L., Pio, B., Pendergrass, T. L., Moyer, M. R., & Patterson, M. D. (2011). Simulation to Assess the Safety of New Healthcare Teams and New Facilities. *Simulation in Healthcare: The Journal of the Society for Simulation in Healthcare, 6*(3), 125–133.

Hollnagel, E. (2009). *The ETTO Principle: Efficiency-Thoroughness Trade-Off: Why Things that Go Right Sometimes Go Wrong.* Farnham, UK: Ashgate Publishing.

Hollnagel, E. (2014). *Safety-I and Safety-II: The Past and Future of Safety Management.* Boca Raton, FL: CRC Press.

Hollnagel, E. (2017). *Safety-II in Practice: Developing the Resilience Potentials.* Abingdon, Oxon: Routledge.

Hollnagel, E. (2018) FRAM – the Functional Resonance Analysis Method for Modelling Non-Trivial Socio-Technical Systems. Retrieved from http://www.functionalresonance.com/. Accessed 30 September 2018.

Iedema, R., Mesman, J., & Carrol, K. (2013). *Visualising Health Care Practice Improvement: Innovation from Within.* Boca Raton, FL: CRC Press.

Kihlgren, P., Spanager, L., & Dieckmann, P. (2015). Investigating Novice Doctors' Reflections in Debriefings after Simulation Scenarios. *Medical Teacher, 37*(5), 437–443.

Kolbe, M., Burtscher, M. J., & Manser, T. (2013). Co-ACT – A Framework for Observing Coordination Behaviour in Acute Care Teams. *BMJ Quality & Safety, 22*(7), 596–605.

Lockman, J. L., Ambardekar, A. P., & Deutsch, E. S. (2015). Optimizing Education with in situ Simulation. In J. C. Palaganas, J. C. Maxworthy, C. A. Epps, & M. E. Mancini (Eds.), *Defining Excellence in Simulation Programs* (pp. 90–98). China: Wolters Kluwer.

Lowe, D. J., Ireland, A. J., Ross, A., & Ker, J. (2016). Exploring Situational Awareness in Emergency Medicine: Developing a Shared Mental Model to Enhance Training and Assessment. *Postgraduate Medical Journal, 92*(1093), 653–658.

Manser, T., Foster, S., Flin, R., & Patey, R. (2013). Team Communication During Patient Handover from the Operating Room: More than Facts and Figures. *Human Factors: The Journal of the Human Factors and Ergonomics Society, 55*(1), 138–156.

Nemeth, C. P., Wears, R. L., Woods, D. D., Hollnagel, E., & Cook, R. I. (2008). Minding the Gaps: Creating Resilience in Healthcare. In K. Henriksen, J. B. Battles, M. A. Keyes, & M. L. Grady (Eds.), *Advances in Patient Safety: New Directions and Alternative Approaches (Performance and Tools)* Volume 3 (pp. 1–13). Rockville, MD: Agency for Healthcare Research and Quality Publication.

Rasmussen, M. B., Dieckmann, P., Barry Issenberg, S., Ostergaard, D., Soreide, E., & Ringsted, C. V. (2013). Long-Term Intended and Unintended Experiences After Advanced Life Support Training. *Resuscitation, 84*(3), 373–377.

Tannenbaum, S. I. & Cerasoli, C. P. (2013). Do Team and Individual Debriefs Enhance Performance? A Meta-Analysis. *Human Factors: The Journal of the Human Factors and Ergonomics Society, 55*(1), 231–245.

Wears, R. L., Hollnagel, E., & Braithwaite, J. (2015). *Resilient Health Care, Volume 2: The Resilience of Everyday Clinical Work.* Abingdon, Oxon: Ashgate Publishing.

Weick, K. E. & Sutcliffe, K. M. (2007). *Managing the Unexpected: Resilient Performance in an Age of Uncertainty,* 2nd ed. San Francisco, CA: Jossey-Bass.

Woods, D. D. (2015). Four Concepts for Resilience and the Implications for the Future of Engineering. *Reliability Engineering & System Safety, 141,* 5–9.

Part IV

Empiricising Boundaries

8

Looking Back Over the Boundaries of Our Systems and Knowledge

Kate Churruca, Janet C. Long, Louise A. Ellis, and Jeffrey Braithwaite
Macquarie University

CONTENTS

Introduction

Resilient health care (RHC) is a young but growing field (Braithwaite, Wears, & Hollnagel, 2015) that can be dated from the formation of the Resilient Health Care Network (RHCN) in 2012. It draws on earlier scholarship and theories, principally resilience engineering studies, as well as disciplines as far afield as cognitive, experimental and social psychology, systems studies, ergonomics, sociology, complexity science, anthropology and philosophy. To date in the literature and across the books assembled by the community of RHC scholars, a rich database of empirical work has emerged (Braithwaite, Wears, & Hollnagel, 2017; Hollnagel, Braithwaite, & Wears, 2013a, 2019; Wears, Hollnagel, & Braithwaite, 2015). Typical in new disciplines, and as befits a social science approach to organisational phenomena, case study accounts dominate this fieldwork.

In this chapter, we present a review of 30 cases of RHC that have been gathered together through the publication of four previous volumes under the auspices of RHCN (Braithwaite et al., 2017; Hollnagel et al., 2013a, 2019; Wears et al., 2015). We applied a classification system to these chapters to identify the settings, the methods used and each chapter's focus. In light of the present volume's focus on *Working Across Boundaries*, we wanted to evaluate this theme in relation to the chapters reviewed.

Boundaries, it seems at first glance, are ubiquitous in health care. They include the professional boundaries between doctors and nurses (Nancarrow & Borthwick, 2005), health care professionals and patients (Griffith & Tengnah, 2013), and the silos between groups, teams and organisations (Braithwaite, 2010). Resilience theories, in a different way, acknowledge the importance of boundaries, noting how health care systems ideally operate within a range of acceptable performance; pressure to reduce costs and staff workloads can, however, mean crossing the boundary into unacceptable performance (Nemeth, Wears, Woods, Hollnagel, & Cook, 2008). Notwithstanding these well-articulated boundaries and theoretical accounts of bounded behaviours and structures, in the analysis that follows, we use a more grounded approach. Our guiding research questions were as follows: what types of boundaries are presented within and across the empirical cases of RHC thus far assembled? How are boundaries, and working across boundaries, related to resilient performance?

Method

A narrative synthesis was undertaken of the 30 case studies we uncovered across the previous volumes in the RHC domain. We first profiled them to understand demographic information, such as country of origin, setting, and participants, and methodological information, such as type of study or focus of study. Our subsequent analysis was guided by an attempt to identify boundaries in these empirical works, and the evidence of working across these boundaries for system resilience.

Results

Overview of Cases

The 30 case studies originated largely from Organisation for Economic Co-operation and Development (OECD) and mainly English-speaking countries, with Canada and the United Kingdom providing the largest number. The methods for studying resilience were primarily qualitative and observational, and the main settings in which research took place were hospitals. Additionally, a mix of clinicians and associated staff most frequently participated in the studies. Although there was overlap in the content of the chapters included in our analysis, we were generally able to classify chapters as having one of three dominant focuses: research striving to *understand everyday clinical work* ($n = 19$); research that attempted to actively *do resilience engineering* ($n = 6$); and research that provided *retrospective analyses of an internal or external disturbance* ($n = 5$). A summary of our profile is presented in Table 8.1.

A number of boundaries were identified within, across and between the cases we reviewed. As we will go on to show, these different types of boundaries have implications for how we understand resilience and study RHC.

Collaboration and Working across Professional and Organisational Boundaries

Many of the chapters illustrated the ways in which working across professional and organisational boundaries contributes to resilient performance. Health care has often been conceptualised as having numerous gaps demarcating the professions, with silos of working that are associated with anthropologically oriented accounts of clinical tribalism (Braithwaite et al., 2013). However, complex problems inherent in everyday clinical work often require multiple viewpoints and collaborative team functioning for resilient system performance. For example, Horsley, Hocking, Julian, Culverwell, and Zijdel's (2019) preliminary evaluation of an intervention in the intensive care unit (ICU) found that teams who facilitate multiple viewpoints produce better situational attunement (anticipate); are more comfortable with raising concerns (monitor); have clearer plans and roles in place (respond) and debrief following events (learning).

Communication among different professions is particularly important during disturbances (Nyssen & Blavier, 2013). In this regard, a study by Ekstedt and Cook (2015) demonstrated how fundamental communication was to adaptive responses by home care health professionals during a blizzard in Sweden. It provided an opportunity for them to make their views explicit, and to share their willingness to make adjustments with colleagues to ensure work was done. Another study by Pariès, Lot, Rome, and Tassaux (2013) highlighted that some recognition of boundaries between professions is potentially useful: 'coopetition' – a mix of cooperation and competition – motivated different groups to succeed in the face of challenging conditions. Other cases, such as one by Heggelund and Wiig (2019) examining maternity settings in Norway, illustrated how flexible learning and a culture of openness could be fostered in support of resilient performance.

TABLE 8.1

Summary of Cases

Demographics	Number of Cases
Country of Study	
Australia	3
Belgium	2
Brazil	1
Canada	5
Denmark	1
Japan	2
New Zealand	2
Norway	2
Sweden	1
Switzerland	1
Taiwan	2
United Kingdom	6
United States	2
Setting	
Hospital	26
Aged care	1
Multiple	3
Participants[a]	
Doctors	2
Doctors and managers	1
Nurses	1
Nurses and doctors	8
Allied health professionals	1
Staff and patients	1
Multiple health care staff[b]	16
Methods	
Qualitative	24
Quantitative	1
Mixed methods	5
Design	
Intervention	5
Observation	25
Focus of Study	
Resilience engineering	6
Understanding everyday clinical work	19
Retrospective analysis of disturbance	5

[a] Classifications of participants are mutually exclusive.

[b] Includes allied health staff, in addition to, for example, doctors and nurses, and administrative staff.

Differing perspectives among organisations and groups, on the other hand, can conspire to impede resilient performance. Laugaland and Aase (2015) conducted a study in Norway on care transitions, an area of care delivery that involves attempts to bridge boundaries between two systems – in this case, primary care and hospital care. They found that one organisation's definition of an 'acceptable' outcome, for instance, might not be the same for another organisation. To avoid these potential limits on cross-boundary resilient, coordination among groups is preferred, as is an explicit sharing and understanding of other stakeholders' perspectives, though usually making these practices commonplace demands time and resources (Braithwaite, Clay-Williams, & Hunte, 2017).

Boundaries Manifesting in Complex Systems

While professional and organisational boundaries exist in many different ways in health care, and multiple studies suggested they can affect resilient performance, there was evidence from the cases that boundaries are to some extent arbitrary – in the sense that they are the product of intricate forces including history, negotiations and trade-offs. In this regard, a number of chapters we reviewed argued for the conceptualisation of health care as a complex adaptive system (CAS) (e.g., Laugaland & Aase 2015; Nakajima & Kitamura, 2019; Sheps, Cardiff, Pelletier, & Robson, 2015). Characteristics of CASs include path dependence, emergent behaviours, relative homeostasis punctuated by intermittent phase transitions, and interdependencies of stakeholders and their artefacts. CASs are composed of multiple, diverse agents (e.g., health professionals, managers, policymakers, patients and other stakeholders) who interact dynamically over time. One prominent and popular example amongst RHC members is emergency departments (EDs) in hospital settings. EDs are seen to exhibit dynamic and non-linear behaviours, changing over time in the face of prevailing conditions, and having a mix of hierarchies, heterarchies, networks and other formal and informal social structures within which agents' actions are constituted (Braithwaite, Wears, & Hollnagel, 2017). As a result, the 'boundaries' of an ED are always shifting (Braithwaite et al., 2013), and behaviours and workloads must constantly be adjusted to account for that.

Due to their variably connected, interacting, interdependent nature, complex systems are relatively open, exhibiting permeable or fuzzy boundaries (Braithwaite et al., 2013). This means, that while professional groups, wards or departments may distinguish themselves from other professional groups, wards and departments, and emphasise their uniqueness, their behaviours are in fact mutually interdependent (Braithwaite et al., 2013). For example, Stephens, Woods, and Patterson (2015) studied how clinical microsystems in the hospital (ED, mental health, intensive care), despite being seemingly bounded, influenced one another's 'capacity for manoeuvre' in the face of disturbances.

The unpredictable nature of complex systems can make anticipation – a resilience potential – more difficult. In another example from the ED, Hunte and Marsden (2019) showed that learning from past events to prepare for future disturbances required not only recognition of patterns in historical data, but multilevel stakeholder buy-in, going beyond ED boundaries. All told, features of CASs in health care, including self-organisation of staff, distinguishable cultures, patterned social behaviours and informal networks appear to be generative of resilient performance (Braithwaite et al., 2013).

Boundaries around Work and between Roles

In the cases we evaluated, boundaries were not always seen as antithetical to resilient practice. There was some evidence that demarcating (i.e., putting boundaries around) aspects of work assisted with maintaining performance during disturbances. For example, Hunte (2015) studied a Canadian ED's response to the large influx of patients following a hockey riot. He identified one of the adaptive responses as the creation of decontamination stations outside to wash out tear gas – the creation of this structural boundary ensured no residue made its way into the ED.

Indeed, despite the potential pitfalls created by emphasising professional boundaries and reinforcing organisational hierarchies, Hunte and Wears (2017) indicated that these aspects of bureaucracy can also be empowering rather than constraining. They particularly noted how work performance can be enabled by guidance and clarity of roles and responsibilities. Clarification of roles and responsibilities requires, to some extent, social structure, entailing a boundary or boundaries between people. Horsley et al. (2019) took up this idea in their intervention to encourage resilient performance, which included a component around the promotion of role clarity between clinicians working in ICU teams in New Zealand. Similarly, in their study, which used simulation to train U.S. staff in a new paediatric community hospital, Deutsch, Fairbanks, and Patterson (2019) demonstrated the importance of clear differentiation of a leadership role. They showed in their study that the leader should only lead and not become involved in providing care.

Boundary-Crossing between Work-as-Imagined and Work-as-Done

Our chapter review further illustrated that, in a very real sense, those engaged in resilient systems work across – or indeed transgress – boundaries constantly, as they overcome both the delicate and fragile, and the more robust and formidable barriers between the idealised ways of working laid out by policymakers and management, and the actual demands of everyday clinical work (Chuang & Wears, 2015). In this sense, there are always multiple small gaps, and sometimes large gulfs, between Work-as-Imagined (WAI) by policymakers, managers and others at the blunt end, and Work-as-Done (WAD) by clinicians at the sharp end, on the front-lines of care. The assumption that care is, or even can be, delivered in standardised ways in accordance with mandated, top-down procedures is misguided, with multiple authors noting how variation and adjustment are part and parcel of health care delivery, and mostly lead to safe and desirable clinical outcomes (Braithwaite et al., 2013; Ross, Anderson, Cox, & Malik, 2019).

Workarounds, which are ubiquitous in health care, manifest at the boundaries between WAI and WAD. These departures from accepted protocols and formalised, prescribed ways of working are useful – they are not only the prime heuristic for getting things done in health care but also provide insights into how people respond to, and overcome, stifling or unworkable policies and procedures, conflicting demands and entrenched barriers in the workplace (Debono et al., 2019). For example, Sujan, Spurgeon, and Cooke (2015) identified a 'secret second handover' in ED, in which paramedics ensured the safety of their patients by transferring necessary yet informal information to treating health care professionals, instead of intermediaries, as prescribed by policy.

Despite being supportive of adaptation in the short term (Debono et al., 2019; Ekstedt & Cook, 2015), some authors cautioned that workarounds may perpetuate gaps in knowledge and limit insights into how everyday clinical work is actually accomplished (Nakajima & Kitamura, 2019). For example, an Australian study by Debono et al. (2019) demonstrated that because workarounds can be mobilised to address immediate, rather than underlying, problems, they can, over time, perpetuate frustration among staff and exacerbate the propensity for brittleness in the system, contributing to the separation of WAI and WAD. In this regard, Hunte and Wears (2017) suggested that more productive attempts to manage the WAI–WAD interface could be useful, including the formation of generative partnerships to forge links among those at the sharp and blunt ends. Resilience can arise from realignment of WAI and WAD was demonstrated by Zhuravsky (2019) in his study of resilient performance in the aftermath of the earthquake in Christchurch, New Zealand.

Boundaries in Thinking about Resilience: A System Property or Action?

Across the chapters in our review, there were a number of theoretical and practical debates about RHC. These debates generally fell into one of two camps regarding the issue of whether resilience is something that can be measured or whether the resilience potentials can be measured (Hollnagel, Braithwaite, & Wears, 2013b). Some like Horsley et al. (2019) and Laugaland and Aase (2015) suggested that resilience is an underlying property of organisations. Yet others, such as Ekstedt and Cook (2015) and Hunte (2015) subscribed more to a view that resilience is something a system *does*, not *has*, a characteristic of how a system performs. Hence, it is only manifestations or expressions of the resilience potentials in the existing conditions, including environmental resources and staff capacity that are amenable to observation.

This boundary in thinking about resilience was evident in the chapters we reviewed, in that some focused on how resilient performance can be strengthened, proposing tools and approaches such as the Functional Resonance Analysis Method (Hounsgaard, Thomsen, Nissen, & Bhanderi, 2019; Sujan & Spurgeon, 2019), the Resilience Markers Framework (Furniss, Robinson, & Cox, 2019), the Concepts for Applying Resilience Engineering model (Anderson et al., 2019) and simulation (Deutsch et al., 2019). Other authors, however, doubted whether resilience can be 'improved', finding no compelling case demonstrating this (Cook & Ekstedt, 2017).

Discussion

In our review of 30 cases of RHC, we found that the majority of the chapters were from English-speaking, OECD member countries and focused on hospital care. They largely used qualitative methods informed by resilience engineering thinking but were also heavily influenced by varied social science traditions. Typically, they were observational rather than interventional, and concentrated on assessing everyday clinical work. Numerous boundaries were identified in the cases we reviewed; these have implications for understanding resilience and for future studies of RHC.

We identified boundaries between professionals and organisations; however, as has been argued by scholars taking a complex systems perspective, such boundaries are recognised as being permeable and shifting rather than solid and fixed. In many of the cases, explicit attempts to acknowledge and work across the flexible and ever-shifting boundaries were conducive to resilient performance. This has been found previously in other empirical work, with Braithwaite (2010) arguing in a systematic review of cross-boundary behaviours in health care that 'identification with one's primary group or profession is very strongly held, yet cross-departmental, cross-group, cross-professional communication, collaboration and interaction are crucial in creating more pluralist, informed and supportive workplaces' (Braithwaite, 2010, p. 330).

We also found evidence to suggest that boundaries were not wholly problematic in terms of resilient performance. In differentiating work tasks, groups, teams and professions, boundaries can give some structured order to a complex system, motivate performance and provide clarity around work responsibilities. Indeed, other research outside of the RHCN indicates that understanding of others' roles and responsibilities is a precursor for successful collaboration in health care (Suter et al., 2009); in this sense, it appears that recognition of interpersonal boundaries provides opportunities for working across them.

Another type of boundary, prominent in many chapters, concerned the difference between WAI and WAD. In working across this potential disjuncture, health care professionals relied upon workarounds. While adaptive in the short-term, some authors of the included chapters raised concerns about the long-term implications of workarounds. When clinicians on the front-lines work around formal prescriptions in secret, for instance, the bigger problem of reconciling WAI and WAD does not get resolved. As a corollary, everyday clinical work remains opaque and poorly understood, which may in fact diminish the capacity for resilient performance in the longer term. Other evidence compiled in a scoping review further underscores the potential detrimental impact of workarounds on patient safety (Debono et al., 2013). As might be expected, there are no easy or universal answers to bridging gaps between WAI and WAD.

We extrapolated the final theme by comparing across chapters, rather than analysing within them. This concerned the boundary in perspectives of chapter authors regarding the nature of resilience itself. From this distinction between resilience as something a system *has,* versus something a system *does,* flowed different assumptions about how to study it, and questions about whether or not it is something that can be subject to interventions or improvements.

Conclusion and Implications for Further Work in RHC

There was a preponderance of high-income countries, hospital-based research and observational studies in the chapters reviewed. Hence, while understanding resilience in everyday clinical work has been a major theme of this body of research, more work is needed with a wider range of mixed methods studies across a diversity of contexts. This would promote greater understanding of this complex phenomenon and improve our potential to enact resilience in everyday clinical work across different health care settings – or at the very least, help us understand RHC and its relationship to boundaries more fully. Even if interventions are attempted and fail, there is much to be discovered about how CASs, such as health care, work and how resilient performance and brittleness in systems show themselves.

References

Anderson, J. E., Ross, A. J., Black, J., Duncan, M., Hopper, A., Snell, P., & Jaye, P. (2019). Resilience Engineering for Quality Improvement: Case Study in a Unit for the Care of Older People. In E. Hollnagel, J. Braithwaite, & R. L. Wears (Eds.), *Delivering Resilient Health Care,* Volume 4 (pp. 32–43). New York, NY: Routledge.

Braithwaite, J. (2010). Between-Group Behaviour in Health Care: Gaps, Edges, Boundaries, Disconnections, Weak Ties, Spaces and Holes. A Systematic Review. *BMC Health Services Research, 10*(1), 330.

Braithwaite, J., Clay-Williams, R., & Hunte, G. S. (2017). Understanding Resilient Clinical Practices in Emergency Department Ecosystems. In J. Braithwaite, R. L. Wears, & E. Hollnagel (Eds.), *Resilient Health Care, Volume 3: Reconciling Work-as-Imagined and Work-as-Done* (pp. 89–102). Boca Raton, FL: Taylor & Francis Group.

Braithwaite, J., Clay-Williams, R., Nugus, P., & Plumb, J. (2013). Healthcare as a Complex Adaptive System. In E. Hollnagel, J. Braithwaite, & R. L. Wears (Eds.), *Resilient Health Care* (pp. 57–73). Farnham, UK: Ashgate Publishing.

Braithwaite, J., Wears, R. L., & Hollnagel, E. (2015). Resilient Health Care: Turning Patient Safety on its Head. *International Journal for Quality in Health Care, 27*(5), 418–420.

Braithwaite, J., Wears, R. L., & Hollnagel, E. (Eds.). (2017). *Resilient Health Care: Reconciling Work-as-Imagined and Work-as-Done*. Boca Raton, FL: Taylor & Francis Group.

Chuang, S. & Wears, R. L. (2015). Strategies to Get Resilience into Everyday Clinical Work. In R. L. Wears, E. Hollnagel, & J. Braithwaite (Eds.), *The Resilience of Everyday Clinical Work* (pp. 225–234). Farnham, UK: Ashgate Publishing.

Cook, R. I. & Ekstedt, M. (2017). Reflections on Resilience: Repertoires and System Features. In J. Braithwaite, R. L. Wears, & E. Hollnagel (Eds.), *Resilient Health Care, Volume 3: Reconciling Work-as-Imagined and Work-as-Done* (pp. 111–118). Boca Raton, FL: Taylor & Francis Group.

Debono, D. S., Clay-Williams, R., Taylor, N., Greenfield, D., Black, D., & Braithwaite, J. (2019). Using Workarounds to Examine Characteristics of Resilience in Action. In E. Hollnagel, J. Braithwaite, & R. L. Wears (Eds.), *Delivering Resilient Health Care* (pp. 44–55). New York, NY: Routledge.

Debono, D. S., Greenfield, D., Travaglia, J. F., Long, J. C., Black, D., Johnson, J., & Braithwaite, J. (2013). Nurses' Workarounds in Acute Healthcare Settings: A Scoping Review. *BMC Health Services Research, 13*(1), 175.

Deutsch, E., Fairbanks, T., & Patterson, M. (2019). Simulation as a Tool to Study Systems and Enhance Resilience. In E. Hollnagel, J. Braithwaite, & R. L. Wears (Eds.), *Delivering Resilient Health Care* (pp. 56–65). New York, NY: Routledge.

Ekstedt, M. & Cook, R. I. (2015). The Stockholm Blizzard of 2012. In R. L. Wears, E. Hollnagel, & J. Braithwaite (Eds.), *The Resilience of Everyday Clinical Work* (pp. 95–74). Farnham, UK: Ashgate Publishing.

Furniss, D., Robinson, M., & Cox, A. (2019). Exploring Resilience Strategies in Anaesthetists' Work: A Case Study Using Interviews and the Resilience Markers Framework (RMF). In E. Hollnagel, J. Braithwaite, & R. L. Wears (Eds.), *Delivering Resilient Health Care*. New York, NY: Routledge.

Griffith, R. & Tengnah, C. (2013). Maintaining Professional Boundaries: Keep Your Distance. *British Journal of Community Nursing, 18*(1), 43–46.

Heggelund, C. & Wiig, S. (2019). Promoting Resilience in the Maternity Services. In E. Hollnagel, J. Braithwaite, & R. L. Wears (Eds.), *Delivering Resilient Health Care* (pp. 80–96). New York, NY: Routledge.

Hollnagel, E., Braithwaite, J., & Wears, R. L. (Eds.). (2013a). *Resilient Health Care*. Farnham, UK: Ashgate Publishing.

Hollnagel, E., Braithwaite, J., & Wears, R. L. (2013b). Epilogue: How to Make Health Care Resilient. In E. Hollnagel, J. Braithwaite, & R. L. Wears (Eds.), *Resilient Health Care* (pp. 227–238). Farnham, UK: Ashgate Publishing.

Hollnagel, E., Braithwaite, J., & Wears, R. L. (Eds.). (2019). *Delivering Resilient Health Care*. New York, NY: Routledge.

Horsley, C., Hocking, C., Julian, K., Culverwell, P., & Zijdel, H. (2019). Team Resilience: Implementing Resilient Healthcare at Middlemore ICU. In E. Hollnagel, J. Braithwaite, & R. L. Wears (Eds.), *Delivering Resilient Health Care* (pp. 97–117). New York, NY: Routledge.

Hounsgaard, J., Thomsen, B., Nissen, U., & Bhanderi, I. (2019). Understanding Normal Work to Improve Quality of Care and Patient Safety in a Spine Center. In E. Hollnagel, J. Braithwaite, & R. L. Wears (Eds.), *Delivering Resilient Health Care* (pp. 118–130). New York, NY: Routledge.

Hunte, G. S. (2015). A Lesson in Resilience: The 2011 Stanley Cup Riot. In R. L. Wears, E. Hollnagel, & J. Braithwaite (Eds.), *The Resilience of Everyday Clinical Work* (pp. 1–10). Farnham, UK: Ashgate Publishing.

Hunte, G. S. & Marsden, J. (2019). Engineering Resilience in an Urban Emergency Department. In E. Hollnagel, J. Braithwaite, & R. L. Wears (Eds.), *Delivering Resilient Health Care* (pp. 131–149). New York, NY: Routledge.

Hunte, G. S. & Wears, R. L. (2017). Power and Resilience in Practice: Fitting a 'Square Peg in a Round Hole' in Everyday Clinical Work. In J. Braithwaite, R. L. Wears, & E. Hollnagel (Eds.), *Reconciling Work-as-Imagined and Work-as-Done* (pp. 119–127). Boca Raton, FL: Taylor & Francis Group.

Laugaland, K. & Aase, K. (2015). The Demands Imposed by a Health Care Reform on Clinical Work in Transitional Care of the Elderly: A Multi-faceted Janus. In R. L. Wears, E. Hollnagel, & J. Braithwaite (Eds.), *The Resilience of Everyday Clinical Work* (pp. 39–58). Farnham, UK: Ashgate Publishing.

Nakajima, K. & Kitamura, H. (2019). Patterns of Adaptive Behaviour and Adjustments in Performance in Response to Authoritative Safety Pressure Regarding the Handling of KCL Concentrate Solutions. In E. Hollnagel, J. Braithwaite, & R. L. Wears (Eds.), *Delivering Resilient Health Care* (pp. 150–159). New York, NY: Routledge.

Nancarrow, S. A. & Borthwick, A. M. (2005). Dynamic Professional Boundaries in the Healthcare Workforce. *Sociology of Health & Illness, 27*(7), 897–919.

Nemeth, C., Wears, R., Woods, D., Hollnagel, E., & Cook, R. (2008). Minding the Gaps: Creating Resilience in Health Care. In K. Henriksen, J. B. Battles, M. A. Keyes, & M. L. Grady (Eds.), *Advances in Patient Safety: New Directions and Alternative Approaches (Vol. 3: Performance and Tools)*. Rockville, MD: Agency for Healthcare Research and Quality.

Nyssen, A. & Blavier, A. (2013). Investigating Expertise, Flexibility and Resilience in Socio-technical Environments: A Case Study in Robotic Surgery. In E. Hollnagel, J. Braithwaite, & R. L. Wears (Eds.), *Resilient Health Care* (pp. 97–110). Farnham, UK: Ashgate Publishing.

Pariès, J., Lot, N., Rome, F., & Tassaux, D. (2013). Resilience in the Intensive Care Units: The HUG Case, in Resilient Health Care. In E. Hollnagel, J. Braithwaite, & R. L. Wears (Eds.), *Resilient Health Care* (pp. 77–96). Farnham, UK: Ashgate Publishing.

Ross, A. J., Anderson, J. E., Cox, A., & Malik, R. (2019). A Case Study of Resilience in Inpatient Diabetes Care. In E. Hollnagel, J. Braithwaite, & R. L. Wears (Eds.), *Delivering Resilient Health Care* (pp. 160–173). New York, NY: Routledge.
Sheps, S., Cardiff, K., Pelletier, E., & Robson, R. (2015). Revealing Resilience Through Critical Incident Narratives: A Way to Move from Safety-I to Safety-II. In R. L. Wears, E. Hollnagel, & J. Braithwaite (Eds.), *The Resilience of Everyday Clinical Work* (pp. 189–206). Farnham, UK: Ashgate Publishing.
Stephens, R. J., Woods, D. D., & Patterson, E. S. (2015). Patient Boarding in the Emergency Department as a Symptom of Complexity-induced Risks. In R. L. Wears, E. Hollnagel, & J. Braithwaite (Eds.), *The Resilience of Everyday Clinical Work* (pp. 129–144). Farnham, UK: Ashgate Publishing.
Sujan, M. A. & Spurgeon, P. (2019). The Safety-II Case: Reconciling the Gap Between WAI and WAD Through Structured Dialogue and Reasoning About Safety. In E. Hollnagel, J. Braithwaite, & R. L. Wears (Eds.), *Delivering Resilient Health Care* (pp. 186–198). New York, NY: Routledge.
Sujan, M. A., Spurgeon, P., & Cooke, M. W. (2015). Translating Tensions in Safe Practices Through Dynamic Trade-offs: The Secret Second Handover. In R. L. Wears, E. Hollnagel, & J. Braithwaite (Eds.), *The Resilience of Everyday Clinical Work* (pp. 11–22). Farnham, UK: Ashgate Publishing.
Suter, E., Arndt, J., Arthur, N., Parboosingh, J., Taylor, E., & Deutschlander, S. (2009). Role Understanding and Effective Communication as Core Competencies for Collaborative Practice. *Journal of Interprofessional Care, 23*(1), 41–51.
Wears, R. L., Hollnagel, E., & Braithwaite, J. (Eds.). (2015). *Resilient Health Care: The Resilience of Everyday Clinical Work*. Farnham, UK: Ashgate Publishing.
Zhuravsky, L. (2019). When Disaster Strikes: Sustained Resilience Performance in an Acute Clinical Setting. In E. Hollnagel, J. Braithwaite, & R. L. Wears (Eds.), *Delivering Resilient Health Care* (pp. 199–209). New York, NY: Routledge.

9

Understanding Medication Dispensing as Done in Real Work Settings – Combining Conceptual Models and an Empirical Approach

Peter Dieckmann
Copenhagen Academy for Medical Education and Simulation (CAMES)

Marianne Hald Clemmensen
Amgros Copenhagen University Hospital

Saadi Lahlou
The London School of Economics and Political Science

CONTENTS

Introduction

Medication dispensation takes place in a socio-technical system, spanning many boundaries in terms of professions, departments and devices (e.g., information systems). Nurses, pharmacy technicians and other health care professionals work in direct or indirect collaboration to identify what medication a patient needs to receive: which dose, which form, via which route and at what point in time. This information is then retrieved from computer systems to identify the correct medication on shelves in a medication room, collect the correct dosage from packages, re-package ready for administering, record the dispensing into the computer system and eventually begin the delivery and administration of the medication to the patients. This work is guided by official rules and regulations, as well as conventions and shortcuts, all of which optimise efforts to reach the ultimate goal – which is to dispense the medication correctly and as quickly as possible. There are widespread connections between the medication room and other parts of the organisation; especially when certain medications are out of stock in the medication room or when new drugs are introduced into the system. All the ingredients for a socio-technical system are there: people, devices and procedures. The activity crosses boundaries and involves several communities and administrative bodies, in addition to technical divisions – therefore, it may be influenced by changes in any of these domains, and conversely, changes in this activity impact these domains. Some influences are direct (e.g., regulations) and some are indirect (e.g., medications that run out of stock).

Literature regarding this domain highlights that this interplay does not function as hoped for in a notable number of cases. Data from national incidence reporting systems in the United Kingdom and Denmark reveal that dispensing errors constitute 12%–18% of all medication errors. Depending on the medication process and research methodology, dispensing error rates are typically reported as being between 0.4% and 4% (Andersen, 2010; Lisby, Nielsen, & Mainz, 2005; Patel et al., 2016; Poon et al., 2006). Visual misinterpretation of names, labels and packaging play an important role in dispensing errors, resulting in confusion between drugs or the incorrect dose being administered (Berman, 2004; Berman, 1969; Cohen, 1994; Cohen, 1993; Cohen, 2002; Emmerton & Rizk, 2012; Hoffman & Proulx, 2003; Schulmeister, 2006). Such errors are potentially harmful to patients (Hoffman & Proulx, 2003; Schulmeister, 2006), and the increasing focus on patient safety and harmful medication errors demands further research on how to reduce the likelihood of visual misinterpretation.

This chapter presents a method to improve our knowledge of the medication process by improving data collection and analysis of dispensation in actual situations; it follows the actors along their paths, within and across boundaries. The method is illustrated by examples from two Danish hospitals, whereby we identify the factors that both facilitate and create barriers to the dispensing of safe medication in the messy conditions of actual health care systems. There are different aspects to consider in the interplay; a method and framework is needed to not only capture relevant snapshots of the system but also – at least as far as this is possible – to make explicit the actual interaction between states, between entities involved and between activities conducted.

Traditionally, safety work tries to understand the 'causes' of the problems and try to remove them from the system. When that is not possible, mechanisms to contain errors should be implemented. The Safety-II approach (Hollnagel, 2014, 2017) sets the focus differently – trying to understand Work-as-Done in real settings. Through a detailed analysis of the affordances in the socio-technical system that guide human action and by analysing the interplay between different parts of the socio-technical system, it is possible to understand the system in such a way that allows us to optimise it and to strengthen good performance – without having to wait until things go wrong. We discuss installation theory (Lahlou, 2017) as a conceptual framework for the description of socio-technical systems and to describe it empirically, we use video-based subject-centred ethnography (Lahlou, Le Bellu, & Boesen-Mariani, 2015) as a novel approach. The theoretical underpinning of installation theory can guide the empirical data collection, analysis and interpretation. We close the chapter with an empirical example to demonstrate the features of this approach.

Installation Theory

Installation theory allows for conceptualising socio-technical systems as a series of units of analysis, installations which are '... specific, local, societal settings where humans are expected to behave in a predictable way. Installations consist of a set of components that simultaneously support and socially control individual behaviour' (Lahlou, 2017). Installation theory is concerned with describing everyday goal-oriented human action in its tension between three layers: the embodied competences of a person (e.g., motivations, emotions, perceptions, cognition, dexterity, knowledge, and skills), social factors (e.g., organisational and social rules, other participants' feedback) and material affordances (e.g., room layouts, availability of resources). The components distributed in the three layers combine to scaffold and channel human action in situ. Because the layers are somewhat redundant, installations are resilient: different layers can compensate for each other. For example, if a certain device is not available for a task, a nurse could retrieve it from another location (physical compensation), or decide to use another procedure (compensation on the rules layer), or try to achieve what the device usually does without it (compensation using embodied competences). Installation theory thus provides a useful lens to describe real work as it unfolds in a systematic manner.

Subject-Centred, Video-Based Ethnography as an Innovative Empirical Method for Data Collection

We can use installation theory to describe the determinants of action as we follow subjects during their activity, using first-person perspective video ethnography. The subjective evidence-based ethnography (SEBE) technique combines two data streams (as seen in Figure 9.1). First, participants

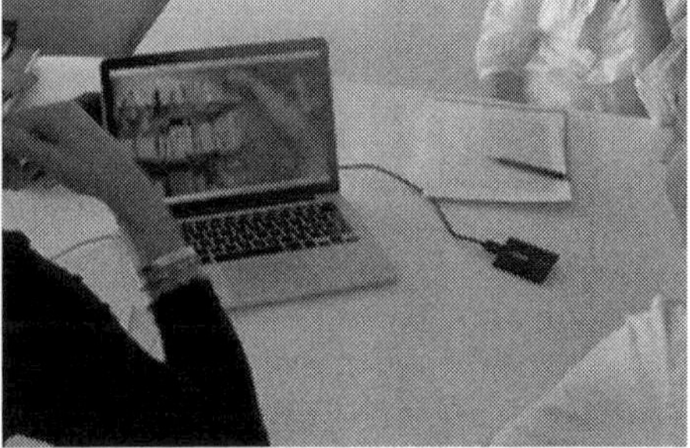

FIGURE 9.1
Two data streams of subject-centred, video-based ethnography (1- left). Recording of Work-as-Done and recording of the replay interview (2- right). (Courtesy of Peter Dieckmann, Marianne Hald Clemmensen, Saadi Lahlou.)

(e.g., nurses) record their own actual work practice with a miniature camera worn on glasses ('subcam'). Each participant reviews their recordings and decides whether they are comfortable for the project team to get the recording. If agreed, the team selects events relevant to the project goals. In a second meeting, the recordings are discussed with the participants, who describe their considerations, feelings and other processes underlying their actions. This second stream of data, the 'replay interview' is also recorded (Figure 9.1). This approach combines the recordings of actions (behaviour) with recordings of the subjective descriptions of the action, and underlying perceptions and considerations. It is important to consider the building of trust and an ethical handling of the data (Dieckmann & Lahlou, in press; Lahlou, Le Bellu, & Boesen-Mariani, 2015). Additional material (e.g., documentation, screenshots, medication packages) can be identified and included in the analysis.

The method has its limitations. Actions and elements outside the view of the camera (e.g., regulations) need to be recalled or drawn into the analysis via different data routes. The replay interview has its own dynamics, and biases can become evident throughout the interviews and during data analysis. To reduce the effect of those biases, it is helpful to make sure that the participants have a detailed understanding of the purpose of the study and the different data streams. Frequent checks on the participants' interpretation of events, and ensuring they are able to reject the researchers' interpretations when they disagree, can counteract biases to some degree.

Case Example

Drug dispensation processes on medical wards in two large hospitals in the capital region of Denmark were recorded. We present stills later to demonstrate the possibilities of the conceptual framework and empirical method. Participants gave informed consent for the publication of the data from the project.

Figure 9.2 shows how the different process steps can be recorded and how the recording perspective provides insight into the Work-as-Done: what steps are taken in which order, which artefacts are used in what way or how different people fulfil the task in similar and yet different ways. The replay interviews provide insights into what perceptions and considerations guide the work: what goals do they try to achieve and what do they try to avoid? Participants describe how their perceptions trigger them to go the next sub-step in the dispensing task, where they gather more information, what information they investigate to find out whether a sub-step is completed and more. Installation theory, with its layers and their interplay, can guide the interview in a systematic fashion. We demonstrate the analytical value of the approach in the following stills.

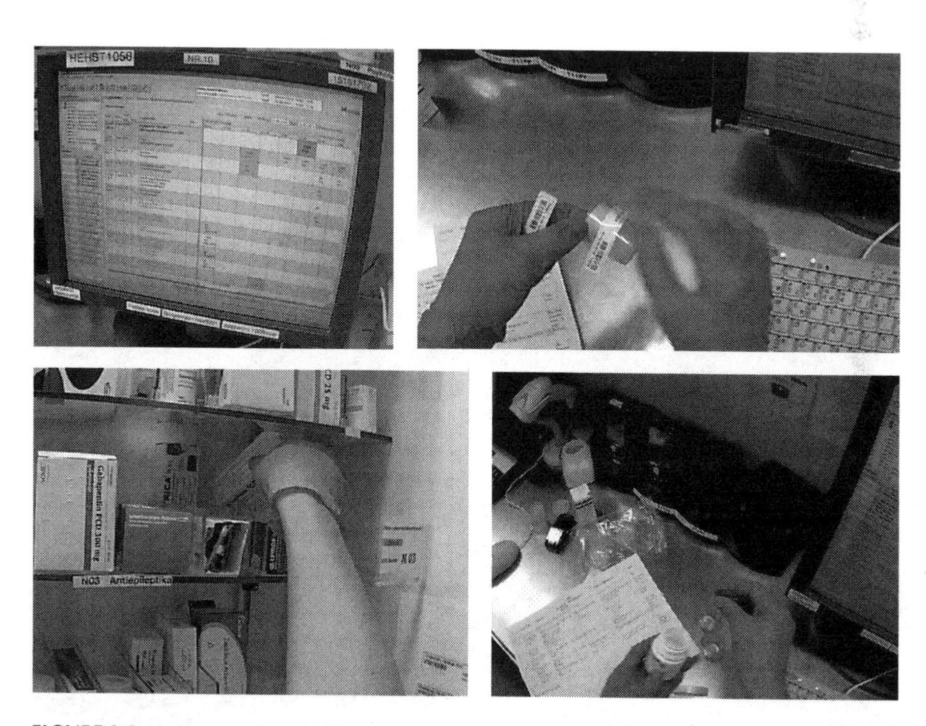

FIGURE 9.2
Preparation of medication for dispensing seen from the nurse's point of view. For each patient, the software (upper left) indicates what drugs must be given. The nurse creates individual containers for a patient with a specific patient barcode label (upper right), gets the medication box from the shelf (lower left), checks it visually then with barcode reader and puts the required dosage in the patient cup (lower right). (Courtesy of Peter Dieckmann, Marianne Hald Clemmensen, Saadi Lahlou.)

Figure 9.3 shows that the actual process involves several information sources and ways to record information (the screen of the software and paper-based lists, serving different purposes). The information is distributed across different parts of the physical layer, and nurses must use their knowledge to utilize the information properly. During the replay interviews, participants describe issues in the software system that make the paper-based instruments necessary, as the functions that the paper fulfils are not implemented in the software. At the organisational level, this indicates which aspects are prioritised in clinical practice. Using a combination of the video recordings and replay interviews to foreground the elements which are taken for granted in daily operations allows them to become the object of conscious reflection. Actors can become more aware of their own competence (and possible gaps in it), they can see how much of their own work is actually intertwined and influenced by decisions made by people and other entities outside of the current context, and they can become aware of how the physical layer impacts their own work in a practical sense.

Figure 9.4 illustrates one way the different layers in an installation can compensate for each other. The medication tray has numbered sections. The numbers correspond to the room (first digit) and the bed's position in that room (second digit): thus, 7–2 denotes bed position 2 in room 7. The tray is optimised for the small medication cups. The latter are colour-coded to denote the period of the day for intake, as per prescription. As some products (e.g., large packaging, infusion bags) do not fit the standard size compartments, the dispensing nurse must find a way of noting and remembering who these are for. Nurses must compensate for the limitations of the material layer with their embodied abilities.

According to procedure, each medication cup should be closed with a lid. This was not done in the featured example (Figure 9.4). While the nurse could be criticised for violating procedures, in the replay interview, the nurse stated that the most important goal was 'dispensing all the medication correctly within 10 minutes'. Recordings show a considerable number of delays that compromise the temporal aspects of the goal. Removing the lid

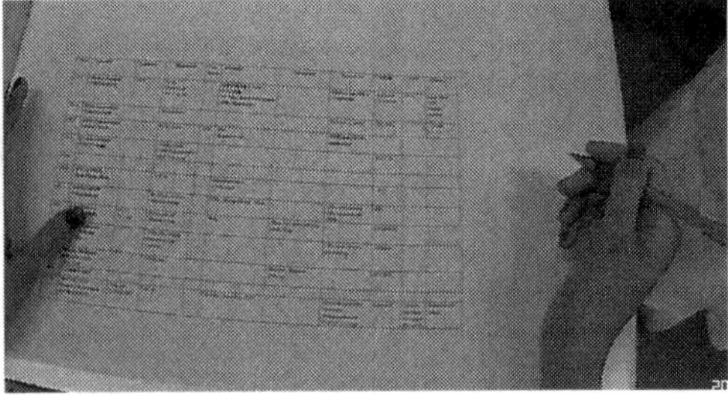

FIGURE 9.3
The nurse–patient log (left). A drug re-order sheet (right). (Courtesy of Peter Dieckmann, Marianne Hald Clemmensen, Saadi Lahlou.)

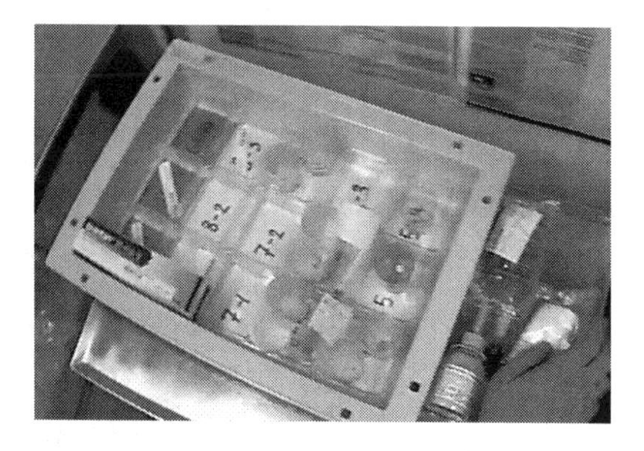

FIGURE 9.4
The medication tray is divided into sections, representing rooms and bed positions (e.g., 7–2). Some of the medications do not fit into the divisions. Note the big box in the upper left and the infusion bags in the bottom. (Courtesy of Peter Dieckmann, Marianne Hald Clemmensen, Saadi Lahlou.)

is a strategy to gain time in the necessary efficiency-thoroughness trade-off (Hollnagel, 2009). The nurse judges how best to achieve the goal. Different priorities and mismatched goals can become clear in the replay interviews as well as trade-offs that need to be made.

Figure 9.5 demonstrates the variability in sub-steps. The different hand positions show how individuals solve similar tasks in different ways, which may correspond to varying efficiency levels. In any case, recordings of separate individuals demonstrate the corridor of normal performance. In many areas of health care, people perform apparently identical tasks in different ways, without this being a conscious reflection. Looking beyond one's own way of doing things, whether individually, as a department, a profession or as a discipline, can help to identify relevant connections across professional and other boundaries. A concrete way of putting this into practice would be to use recordings of practice in group settings, where they can trigger valuable insights about colleagues' work (Iedema, Mesman, & Carroll, 2013).

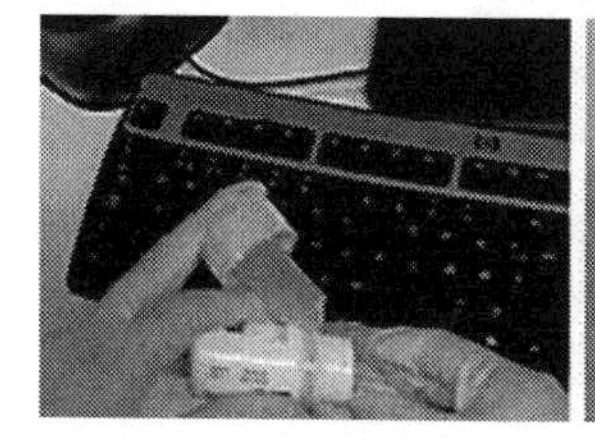
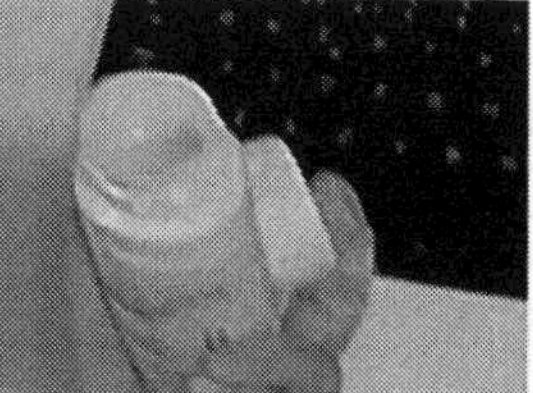

FIGURE 9.5
Illustration of variability in everyday practice. (Courtesy of Peter Dieckmann, Marianne Hald Clemmensen, Saadi Lahlou.)

Discussion and Conclusion

The SEBE method allows for a realistic view on Work-as-Done in actual work systems. The SEBE technique can help to identify relevant aspects of the layers of the installation, their compensation mechanisms and the regular variability in everyday work. Combining the different data streams, the subcam recording and the recording of the replay interview synthesises the action description along with the underlying perceptions and frames. Use of this technique allows varying approaches to similar tasks to be identified, highlights the different elements of the socio-technical system and reveals the compensation mechanisms between the different layers of the installations. The SEBE method can also help to pinpoint the weaknesses in the installation and reveal how subjects manage these limitations in actual practice.

However, the SEBE method also has its limitations. For example, not all relevant elements are seen in the video and might thus be missed during the replay interviews, the recorded performance of a person may or may not be a representative example of the person's work and the recording might change some aspects of practice. However, SEBE's capacity to create an effective stimulation of episodic memories and thus detailed accounts of cognitive processes has great potential, in that it allows for detailed insights into participants' perspectives that would be difficult to capture from any other angle and triggers reflexivity on the Work-as-Done. In addition, it enables participants to observe the work of their colleagues across professional boundaries.

In the case examined, it became clear that many different systems are needed to make the dispensing work. It is essential that nurses handle those systems across the different levels of the installation and constantly compensate for shortcomings on the material and organisational level. The variety of approaches on the individual level could be demonstrated, for example, by examining the physical dexterity involved when handling medication containers. The method also enabled us to make explicit the partly conflicting goals of the nurses' work (time vs. safety), and their efforts to solve the contradictions in their goals. The SEBE method has the propensity to cross multiple real and imagined boundaries in the health care setting, both on the individual and collective levels.

References

Andersen, S. E. (2010). Drug Dispensing Errors in a Ward Stock System. *Basic & Clinical Pharmacology & Toxicology, 106*(2):100–105.

Berman, A. (2004). Reducing Medication Errors Through Naming, Labeling, and Packaging. *Journal of Medical Systems, 28*(1):9–29.

Berman, R. S. (1969). New Methods of Packaging Drugs Help Reduce Medication Errors. *Modern Nursing Home, 23*(6):50–57.

Cohen, M. (1994). Medication Errors: Versed Packaging--Newer Isn't Better. *Nursing,* 24(7):19.

Cohen, M. R. (1993). Drug Alert: Packaging Leads to Fatal Errors. *Nursing, 23*(10):17.

Cohen, M. R. (2002). Medication Naming, Labeling, and Packaging. *American Journal of Health-System Pharmacy, 59*(9):876–877.

Dieckmann, P. & Lahlou, S. (in press). Visual Methods in Simulation-Based Research. In D. Nestel, J. Hui, K. Kunkler, et al. (Eds.), *Healthcare Simulation Research: A Practical Guide.* Cham, Switzerland: Springer.

Emmerton, L. M. & Rizk, M. F. (2012). Look-alike and Sound-alike Medicines: Risks and 'Solutions'. *International Journal of Clinical Pharmacy, 34*(1):4–8.

Hoffman, J. M. & Proulx, S. M. (2003). Medication Errors Caused by Confusion of Drug Names. *Drug Safety, 26*(7):445–452.

Hollnagel, E. (2009). *The ETTO Principle: Efficiency-Thoroughness Trade-Off: Why Things That Go Right Sometimes Go Wrong.* Burlington, VT: Ashgate Publishing.

Hollnagel, E. (2014). *Safety-I and Safety-II: The Past and Future of Safety Management.* Farnham, UK: Ashgate Publishing.

Hollnagel, E. (2017). *Safety-II in Practice: Developing the Resilience Potentials.* New York, NY: Routledge.

Iedema, R., Mesman, J., & Carroll, K. (2013). *Visualising Health Care Practice Improvement: Innovation from Within.* London, UK: Radcliffe Publishing.

Lahlou, S. (2017). *Installation Theory. The Societal Construction and Regulation of Behaviour.* Cambridge, UK: Cambridge University Press.

Lahlou, S., Le Bellu, S., & Boesen-Mariani, S. (2015). Subjective Evidence Based Ethnography: Method and Applications. *Integrative Psychological and Behavioral Science, 49*(2):216–238.

Lisby, M., Nielsen, L. P., & Mainz, J. (2005). Errors in the Medication Process: Frequency, Type, and Potential Clinical Consequences. *International Journal for Quality in Health Care, 17*(1):15–22.

Patel, N., Desai, M., Shah, S., & Ghandi, A. (2016). A Study of Medication Errors in a Tertiary Care Hospital. *Perspectives in Clinical Research, 7*(4):168–173.

Poon, E. G., Cina, J. L., Churchill, W., Patel, N., Featherstone, E., Rothschild, J. M., … Gandhi, T. K. (2006). Medication Dispensing Errors and Potential Adverse Drug Events Before and After Implementing Bar Code Technology in the Pharmacy. *Annals of Internal Medicine, 145*(6):426–434.

Schulmeister, L. (2006). Look-alike, Sound-alike Oncology Medications. *Clinical Journal of Oncology Nursing, 10*(1):35–41.

10

Resilient Front-Line Management of the Operating Room Floor: The Role of Boundaries and Coordination

Sudeep Hegde
University at Buffalo

Cullen Jackson
Harvard Medical School

CONTENTS

System Description

The 'perioperative' environment is a term used for an integrated care system, consisting of units for the three primary phases of surgery: pre-operative (prior to surgery), intra-operative (during surgery) and post-operative (following surgery). Each phase involves coordination of tasks, resources and information between multiple caregiver groups, including surgeons, anaesthesiologists, nurses, nurse anaesthetists and residents. The perioperative environment has an inherent dynamic variability, often reflected in changes in case volume, staff availability and case acuities. Accordingly, the overall schedule of the operating rooms (ORs) needs to be monitored, and adjusted or manipulated, and the limited staffing resources need to coordinate accordingly across the OR floor. Anaesthesiologists, in particular, are shared resources, in that one anaesthesiologist may be assigned to as many as three ORs simultaneously, each with a supervised anaesthesia resident or a nurse anaesthetist. In that sense, their availability is a limiting factor for OR teams in terms of progressing from case to case or for scheduling additional cases during the day.

Role of the OR Floor Manager

Central to the management of resources and schedules on the perioperative care front is the role of the OR floor manager (FM). The FM is typically an anaesthesiologist in charge of coordinating anaesthesia resources between multiple rooms across the OR floor. FMs make decisions related to manoeuvring staff assignments, resource distribution and schedule adjustments, including scheduling 'add-ons' (cases added to the 'waiting list' on the day of surgery or the previous day).

Although primarily concerned with anaesthesia staff, FMs' decisions affect surgery and nursing as well. This is because FMs have the ability to observe, from their vantage point as overseers of the general flow of operations across the perioperative system, the overall system state, including constraints and opportunities for leverage, progress of cases and system capabilities. They are therefore in a relatively advantageous position to coordinate across the physical, functional and hierarchical boundaries that are characteristic of the perioperative environment. In addition to managing the OR floor, FMs are assigned to ORs as attending supervising residents or nurse anaesthetists. This makes the FM's role all the more challenging, as their attention needs to be divided between monitoring and responding to needs across the floor, and attending to their own patients, as seen in Table 10.1.

TABLE 10.1

Tasks and Priorities of OR FMs

Task/Priorities by Phase of Day	Task/Priorities Generic to Phase of Day
Pre-start/Start (6:45 am–7:30 am) • 'Handover' from overnight attending • Ensure on-time starts for all cases Morning (7:30 am–11:00 am) • Cover staff breaks • Maximal utilisation: 'pack the ORs' Afternoon (12:00 pm–3:30 pm) • Cover staff lunch breaks • Plan for fewer ORs and nursing staff End of day-shift (3:30 pm–5:00 pm) • Cover staff breaks • Ensure staff get to leave on time • Handover to oncoming late attending	• Schedule add-on cases (and assign them to rooms) • Get the maximum number of cases done in the shortest amount of time • Manage staff workload – provide breaks; ensure they are not stuck late • Coordinate anaesthesia staff across cases • Provide clinical assistance • Cover sub-specialties with appropriate staff • Respond to emergencies

General Description of Communication Infrastructure and Patterns on the OR Floor

The FM must communicate and coordinate across multiple caregiver groups and various locations in the perioperative environment (as seen in Figure 10.1). A central tool used in monitoring the progress of cases is the OR scheduling board, which exists both in an electronic and handwritten format. The schedule board lists all the cases that are currently scheduled

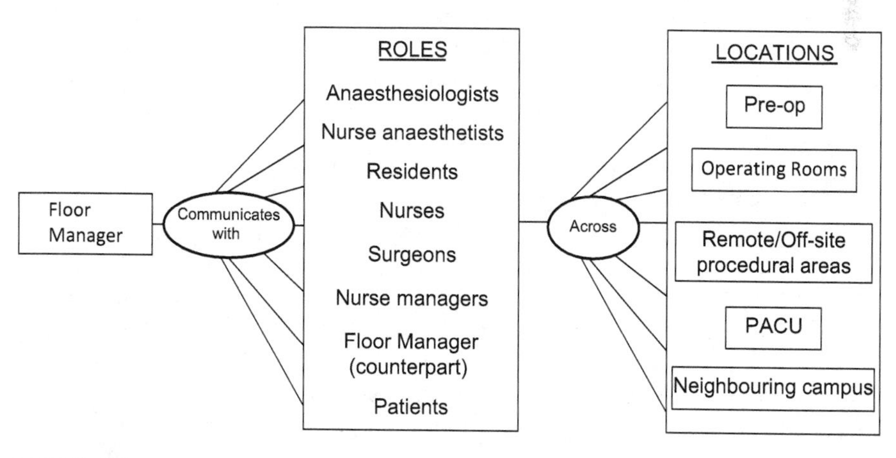

FIGURE 10.1
Various caregiver groups that the FM communicates with and coordinates across various locations.

and waiting to be scheduled, including the start and (expected) end times, and the names of surgical, nursing and anaesthesia staff for each case. The board is continuously updated to reflect changes in case durations, schedule (e.g., cancellations) and staff assignments. All changes made by the FM, such as switching staff between cases, and inserting add-ons, are also reflected on the board. Consequently, the scheduling board serves as a common point of reference for all stakeholders in the perioperative system. For instance, the preoperative nurse can determine when to have the next patient-to-follow ready for surgery, depending on the progress of the current case in the assigned OR. However, the schedule board is far from a complete reflection of the dynamic state of the perioperative system. In reality, the FM is at the centre of a dense web of communication – seeking, updating and transferring information across boundaries, including those of function (e.g., clinical, administrative), location (e.g., OR, post-anaesthesia care unit, holding area, other campus) and time (e.g., case durations, end of shift times). It is through near-constant communication that the FM monitors the ebb and flow of operations and coordinates resources.

The Study

A study was conducted at a large academic multi-specialty hospital, with two adjacent campuses, 'East' and 'West', each with about 25 anaesthesia locations (main ORs and procedure areas). The East campus handles a higher number of cases of shorter duration and lower acuity than the West campus, which has fewer cases of longer duration and higher acuity. Each campus has its own FM, and there is a different FM assigned each day during the work week. The two FMs are scheduled for each day, weeks in advance, from a pool of 18 anaesthesiologists chosen by the department to serve as FMs.

The lead author conducted 15 observation cum-interviews with FMs. The selection of FMs for the study was opportunistic – based on the clinical schedule – in an attempt to conduct observations with as many FMs from the pool as possible and to have a mix of observations on both campuses. A few FMs work on both campuses, and we tried to collect observations on both campuses for them if possible. The focus of our data gathering was to understand the goals and objectives of FMs, the workflows and information flows involved, the typical challenges and concerns encountered and the strategies used to achieve their goals. Furthermore, we conducted interviews and focus groups with perioperative staff to learn 'their side of the story' in terms of how their workflows are influenced by the FM's role. This group included anaesthesiologists, residents and nurse anaesthetists. The experience of the FM participants ranged from 1 month to more than

40 years, with a median of 15 years. Content analysis of data was carried out to identify emergent themes related to challenges to and enablers of effective care delivery from a floor management standpoint, including strategies for monitoring, anticipation and coordination across the various parts of the perioperative environment.

Findings

The following section illustrates a scenario to highlight themes that are discussed subsequently.

SCENARIO

7:00 AM: Start of shift – FM for West campus reviews the schedule of ORs and resources for the day. One of the anaesthesiologists, Anne, has requested an early reprieve owing to a minor back injury. All ORs have cases scheduled. There is just one 'free' anaesthesiologist who is not assigned to any OR, and is available as a back-up, e.g., for covering breaks.

10:13 AM: A transesophageal echocardiography (TEE) case is waiting to be done at a remote location in the hospital. It is non-urgent and expected to be of short duration. Although the FM can potentially send the 'free' attending for the TEE case, he refrains from doing so in order to use the attending to give breaks to other anaesthesia staff on the OR floor.

10:44 AM: East-FM pages the West-FM for help – needs a back-up resource. FM has no extra back-ups today and calls back to convey that he is not able to help at this time.

1:05 PM: The anaesthesiologist, Anne, is finished with her case early in OR-6. In order to allow her to leave early, the FM had planned for the team in OR-12, who were expected to finish by noon, to take over her room after she would leave. This happens as planned/expected. Since the TEE case is expected to be short in duration and non-strenuous, the FM assigns Anne to the case. He calls and intimates the TEE set-up staff. However, it turns out that the case already got done without an anaesthesiologist. Mindful of his East-counterpart's request for back-up earlier, the FM requests Anne to go over and help on the East campus instead. He further calls the East-FM to let him know that he is sending an attending for back-up, but she has to be reprieved quickly given her back problem. The East-FM gratefully agrees to do so.

Insights: Boundaries as a Resource, Boundaries as a Constraint

Several boundaries are represented in the earlier scenario – all of which play an important role in influencing the strategies and means used by the FM in managing the situations described. We identify some categories of boundaries represented here and discuss them in situations beyond this specific scenario. In some situations, boundaries represent constraints that the FM has to plan for and manoeuvre resources to accommodate. In other circumstances, the boundaries are a way of organising the coordination of functions and resources, as well as maintaining order. In some cases, the FM may even create new, albeit temporary, boundaries to achieve certain priorities. We discuss boundaries, identified among the themes generated through the content analysis, in terms of the varied roles they play in resilient floor management.

Functional Boundaries

The FM in this scenario held off on committing his free anaesthesiologist to the remote location for a non-urgent case, thus ensuring availability for supporting other anaesthesia needs on the floor, where the majority of cases are conducted and where any exigencies could immediately strain the available resources. This is an example of the FM creating a boundary for resource allocation to reserve buffers in anticipation of increased workload. Alternatively, this could be interpreted as evidence of the FM's awareness of boundaries later in time and thereby taking anticipatory action by reserving a key resource. Creating this boundary also enabled the FM to prioritise workload management of anaesthesia staff on the OR floor by using his backup attending to give breaks or support for caregivers involved in long/high-acuity cases.

On the other hand, boundaries can constrain resources and flexibility. For instance, urgent or high-acuity cases, such as a liver transplant, which require specialised anaesthesia staff, represent functional limitations that demand allocation of the appropriate resources to the case. The FM, therefore, must prioritise availability of the case-specific specialist caregiver. This in turn creates constraints on assignment of staff to other cases, and generally determines which cases can be done when. Staff fairly frequently call in sick or become unavailable due to health, personal or other reasons. Therefore, although not entirely unexpected, such potential functional limitations need to be factored into the FM's plans or adaptive strategies. In the scenario, the FM had to contend with the limited availability of one particular attending, Anne, because of her back injury. However, in this case, the FM was able to 'stretch' this functional boundary by sending her over to help on the East for a short while, while also remaining empathetic to her condition by informing his East counterpart that she should be reprieved early.

Temporal Boundaries

The day's schedule of cases, and the assignment of anaesthesia staff to the scheduled cases (including shift times), is arranged prior to the day of surgery. These are the primary constraints which the FM has to be aware of in coordinating schedule changes and staff re-assignments. However, it is expected that these boundaries can vary considerably through the course of the day across multiple locations, cases and staff, adding to the complexity of the perioperative environment. The shifting nature of these boundaries creates both constraints and opportunities for the FM. For instance, cases that finish earlier than planned free up staff who can then be used in areas where additional help may be needed, or to provide breaks and ease pressure on colleagues working on more physically and mentally demanding cases. In the scenario, the attending, Anne, finished her case in OR6 early enough that the West FM was able to offer her as a resource to the East FM. However, when cases end later than scheduled, the competition for resources among later or unscheduled cases intensifies, even as available resources are stretched toward their capacities. This, in turn, has a negative effect on the productivity of the perioperative system, as fewer add-on cases can be scheduled during the day. A strategy commonly employed by FMs to counter this is to aggressively schedule add-ons earlier in the day when there is less uncertainty and more resources (in terms of staffing, open rooms) are available. Avoiding bottlenecks later in the day preserves flexibility despite the diminished resources, thereby retaining the capacity to schedule more ad hoc cases or respond to any contingencies.

While it is ideal to complete all cases within the shift they were scheduled for, it is almost a daily occurrence that at least some cases 'spill' over the regular 10- or 12-hour shift of staff, into the evening shift period. Given this expectation, the 'blunt end' of the system has well-developed approaches that 'stagger' temporal boundaries to enhance the certainty of staff availability beyond regular day-shift times. One such system is the assignment of 'late staff' in which a few anaesthesia attendings on the day shift are chosen to be available for cases running later than the regular shift time (usually 5 pm). The late staff are designated based on a tiered framework, as Late 1, Late 2, Late 3 and so on, to indicate the order in which each gets to be released off duty for the day. For example, in the event that two attendings are required for a late case, Late 2 would be released only after Late 1 is released. This system of staggering a temporal boundary (attendings' shift times) provides systemic resource buffers for the FM to utilise against potential case delays, especially towards the end of the day shift.

Physical Boundaries

The scenario involves two campuses – East and West – which are in different buildings almost two blocks from each other. Given his bird's eye view of the system and awareness of individual constraints and priorities, this FM opportunistically leveraged a serendipitous resource, Anne, to respond to a colleague's (East FM) need. Second, within the West campus, the transesophageal echo test case is to be done in a room at a remote location, on a different floor, from the OR floor. Additionally, while the ORs comprise the majority of anaesthesia locations, there are several other locations on other floors, such as radiology and the electrophysiology lab, which must also be staffed by anaesthesiologists. The scenario comprises examples that illustrate the importance of the FM's role in communicating and coordinating resources across multiple locations spread across physical space. The FM's global monitoring and coordinating function ensures there is greater fluidity of resources between the various parts of the perioperative environment that would otherwise be insulated from each other due to physical boundaries and distances.

Hierarchical Boundaries

The FM fulfils a unique role of reconciling long-term, policy-based criteria defined by the hospital's senior management (the blunt end) and the short-term (or immediate) needs of the front-line operations (the sharp end). Blunt-end criteria include OR utilisation and efficiency (e.g., case turnaround times), volume of cases to be completed each day and various quality metrics (e.g., minimising preoperative wait times). The FM accomplishes this by using affordances in the environment, including functional and temporal boundaries described earlier. However, given the complexity of the perioperative system, the FM's role should be seen more as an influencer rather than a controller of operations on the floor.

Perspectives from Perioperative Staff

We held interviews and focus groups to gain insight into the perspectives of perioperative caregivers in terms of their interactions with FMs, and to find out how their workflows are affected by those interactions. Some key insights are as follows:

Reliance on FMs for Global Context

Because caregivers in the ORs or procedural areas are focused on the cases in their specific locations, they don't have a full sense of the global constraints and priorities across the OR floor. This limits their ability to anticipate delays or schedule changes, or to coordinate amongst themselves for support. They therefore rely on the FM, who is expected to maintain a bird's eye view of the entire floor, to make decisions that might affect disparate parts of the perioperative system. For instance, an attending who needs a replacement while dealing with an emergency in another area would have to defer to the FM to determine who might be available.

FM as a Clinical Resource

FMs are anaesthesiologists themselves, so are often seen by caregivers as a resource. They may be called upon by caregivers to provide clinical assistance or a second opinion on a specific patient's anaesthesia care. In that sense, FMs are natural resource buffers who may, for instance, provide cover for a colleague responding to a contingency on another case. For the FM, however, this represents a trade-off between local (OR-level), clinical and global (floor-level) coordination goals, as being constrained in a single clinical location can hamper their ability to monitor other areas and respond in a timely manner. While many experienced FMs offer their services as a final option when no other anaesthesia staff are available for clinical engagement, they must weigh the need to be in an OR against overseeing the floor.

Prompt Communication of Plans

In a dynamic perioperative environment such as that described in the scenario, the earlier individual caregivers are notified about anticipated changes to their schedule, and they are better prepared they are to make any necessary changes and adjust their individual workflow. To illustrate the importance of communicating expectations early, a nurse anaesthetist explained: *'when a plan is communicated (promptly and early), it not only allows us to mentally prepare, but also from a patient safety perspective, gives time to assess the patient. More than anything else, it tells you that you are respected'*. Conversely, when communication is delayed, it shifts the onus towards the point of care, forcing them to adapt under time pressure, potentially narrowing the margin of safety.

Balancing Multiple Priorities across the Floor

Multiple priorities across various perioperative groups can often create competing pressures or potential conflicts, calling into play the FM's role as a mediator. For instance, it is commonly observed that surgeons try to maximise the number of surgeries performed each day. This drive for efficiency can stress other staff in the ORs, including anaesthetists. In such cases, the FM intervenes to find a way to reconcile conflicting goals – perhaps finding a replacement for fatigued anaesthesia staff, or shielding them from a surgeon's pressure by ruling that additional cases cannot be scheduled. When different groups in different locations strive to optimise resources to meet their 'local' goals, the FM can serve as a regulator by either supporting or restricting those efforts in keeping with more 'global' goals.

Implications for Enhancing Resilience in Floor Management

Resilience in OR floor management involves being able to avoid scheduling bottlenecks, and maintaining enough resources and adaptive capacities to be responsive to staff and patient care needs as they arise. This involves active monitoring, anticipation, planning and decision-making, all of which are executed through communication with various groups across the perioperative environment. It is possible that a seemingly 'normal' floor state can quickly turn into a complex exigent state, with multiple emergencies and resource shortages. In such circumstances, the FM's role in adaptive response is highlighted. Less visible, however, are the 'upstream' measures, like resource buffering, which underlie 'normal' or 'in-control' functioning (Ekstedt & Cook, 2015). In a complex network of people, processes, policies and artefacts, separated by boundaries, resilient potentials are mobilised via communication and coordination. This implies that communication channels between individuals and teams, and across varying timescales, need to be open, operational and effective. In front-line OR floor management, this could mean, for instance, enabling preoperative nurses to promptly update the FM about potential delays in patient readiness from the patient's bedside, rather than having to wait until they can leave the patient.

The chapter has identified various boundary categories and their varied roles in front-line perioperative management. While some boundaries may be more rigid, such as the clinical responsibilities of various specialists, others are more ephemeral in nature. The latter include short-term emergent constraints and resources (e.g., cancellation of a case, moving a pending case 'up' the schedule). Given a constant mix of both expected and unexpected boundaries, it is important for the FM to maintain a degree of fluidity of available resources – being able to re-allocate, manoeuvre or adjust them as needed. As demonstrated in the scenario, FMs strive to work around existing boundaries or create new ones to meet the needs of the clinical environment. Furthermore, boundaries serve as a cognitive scaffolding for FMs in the management of ORs, in that they serve as anchors for planning, decision-making, and coordinating (Clark, 1998). The FM's level of awareness of these boundaries is crucial to their effectiveness as facilitators and coordinators of resources and functions across the perioperative system.

Although this chapter describes an individual role in the context of resilient health care, it is not meant to be a study of individual resilience. Rather, floor management is a systemic function (or set of functions) centred around an individual. This study abstracts key aspects of the FM's role, which can be used to inform design of system-level support or policies for resilient floor management in general. With increasing patient volumes and complexity of case acuities, hospitals are under increased pressure to meet efficiency demands and keep costs low, while precariously maintaining a reasonable margin of safety (McGinnis, Stuckhardt, Saunders, & Smith, 2013). Therefore, further directions for this work could include a study of factors that constrain or enable efficient exchange of information between perioperative teams and FM. Insights on front-line barriers and enablers of add-on scheduling and staff assignments can be used to develop a framework to inform OR scheduling practices. The aim of such an effort could be to minimise the possibility of schedule conflicts, or conversely, to maximise the ability of the perioperative system to create buffers or backups when needed.

Conclusion

From a systems standpoint, the FM's role is an important one that is meant to bridge gaps in communication and coordination between disparate teams, locations and units within the perioperative environment. Perioperative management may also be viewed as coordination across various types of systemic boundaries, such as functional, temporal, physical and hierarchical. FMs also function as natural backup resources that can be easily accessed by anaesthesia staff whenever help is needed. In making scheduling and staff coordination decisions, and in negotiating with various caregiver groups, they influence trade-offs between competing goals such as efficiency, throughput and safety. From a resilience standpoint, there is a need for further study on how communication and coordination in everyday OR floor management can be better supported systemically.

Acknowledgements

We thank the staff in the Department of Anaesthesia, Critical Care and Pain Medicine at Beth Israel Deaconess Medical Center (Boston, MA) for their participation in the study. We are especially grateful to senior anaesthesiologists, Peter Panzica, MD, and Philip Hess, MD, and Chief Nurse Anaesthetist, Beth Coolidge, CRNA, for facilitating our data collection efforts and providing insights crucial to the design of the study and assimilation of data.

References

Clark, A. (1998). *Being There: Putting Brain, Body, and World Together Again*. Cambridge, MA: MIT Press.

Ekstedt, M. & Cook, R. (2015). The Stockholm Blizzard of 2012. In R. Wears, E. Hollnagel, & J. Braithwaite (Eds.), *Resilient Health Care, Volume 2: The Resilience of Everyday Clinical Work* (pp. 59–74). Farnham, UK: Ashgate Publishing.

McGinnis, J. M., Stuckhardt, L., Saunders, R., & Smith, M. (Eds.). (2013). Imperative: Managing Rapidly Increasing Complexity. In *Best Care at Lower Cost: The Path to Continuously Learning Health Care in America*. Washington, DC: National Academies Press.

11

Patient Flow Management: Codified and Opportunistic Escalation Actions

Jonathan Back and Janet E. Anderson
King's College London

Alastair J. Ross
University of Glasgow
King's College London

Peter Jaye and Katherine Henderson
Guy's and St. Thomas's NHS Foundation Trust

CONTENTS

Introduction

Patient flow is the movement of patients through stages of care within a hospital, and as such requires effective working across boundaries. Each stage of care requires coordination across an organisational boundary and can involve moving into a different area of the hospital with different challenges, performance targets and standards of care that need to be met. The flow of patients through the emergency department is an example of patient flow that can be problematic if free beds are not available in other wards due to flow difficulties. In this chapter, we will outline the prerequisites for an effective response to patient flow pressures and discuss two case studies of work practices that have evolved to manage flow across organisational boundaries. An argument will then be presented calling for a better coordinated, more effective and sustainable approach to patient flow transitions across boundaries.

We reflect on our work at a large National Health Service (NHS) hospital in England, conducted by the Centre for Applied Resilience in Healthcare (*see* protocol by Anderson et al., 2016a). Resilience engineering directs attention to studying how health care work is carried out in practice, because work does not always fit pre-specified actions (e.g., Wears, Hollnagel, & Braithwaite, 2015). Instead, health care work requires adaptive capacity, which involves staff managing pressures and problems by making in situ adaptations and goal trade-offs to achieve desirable outcomes (Hoffman & Woods, 2011). By studying these adaptations, it is thought that we can strengthen good practice (Hollnagel, Pariès, Woods, & Wreathall, 2011).

Background

What Is Escalation?

Across the NHS, demand for hospital care often exceeds capacity, leading to patient flow pressures that manifest in the emergency department. Escalation is the process of identifying when demand is increasing, hence intensifying efforts, so that performance targets and standards of care can continue to be met. This can be attempted in a number of ways, such as reducing the number of admissions and expediting discharges from hospital. The focus is on optimising throughput times within the emergency department and inpatient wards. Escalation requires working across organisational boundaries, as a hospital-wide response is needed in order to be effective, according to the NHS conceptualisation (NHS Improvement, 2017). Although escalations are initiated by a department or unit in a hospital, resources will sometimes be garnered from elsewhere in the system for the escalation to be successful.

NHS Escalation Policies

All NHS hospital trusts have escalation policies that specify when escalation should happen and the required responses. These policies are triggered when normal activities need to be changed in response to actual or anticipated patient flow pressures. Depending on the state of the system, different escalation levels are proposed. These levels include:

- Green – where no action is needed;
- Amber – where existing resources need to be reconfigured;
- Red – where additional resources need to be garnered;
- And, black – where crisis management is invoked.

Despite the prolific use of escalation policies across the NHS, there is a surprising lack of evidence for their effect on organisational or patient outcomes. Moreover, policies have been developed in a somewhat ad hoc manner, with no standard process for development or implementation. Escalation policies epitomise 'Work-as-Imagined'. This term refers to the various assumptions, explicit or implicit, that people have about how their own or others' work should be done (Dekker, 2006; Ombredane, & Faverge, 1955).

Codified Escalation Actions in Emergency Departments

Our recent study of escalation focused on the use of escalation actions by emergency departments (Back et al., 2017). These actions are codified in policy and are designed to enable emergency departments to manage increases in demand (e.g., a sudden inflow of patients) or handle a reduction in capacity (e.g., a lack of beds in the hospital system to admit patients). Escalation policies used by 20 different NHS hospital trusts were reviewed. It was found that there was an implicit assumption that there is spare capacity in the system which can be called into action and quickly reconfigured to ease pressure. For example, common actions specified in policies involved reassigning clinical staff who are performing administrative/training roles to clinical roles, or redesignating bed/cubicle usage in a less pressured area for use by a busy area. However, in an under-resourced NHS, these spare capacities are eroded as part of everyday work, and the capacity for resilient performance is therefore reduced, even when specified by policies. Our ethnographic study of escalation in practice, conducted at a large NHS hospital in England, found that codified actions could often not be implemented because there was no spare capacity. Moreover, actions recommended by policies to address flow bottlenecks and relocate patients were rarely successful. For example, streaming patients at the front door without a thorough assessment could result in patients being sent to the wrong area of the emergency department, increasing workload and potentially jeopardising care. Although actions are specified by escalation policies, they are often devised as a way of behaving in response to a predicted situation, established before the event. The predetermined resources that codified actions require to function were often absent or depleted, leading to a brittle rather than resilient system.

Margins for Manoeuvre

It could be argued that in the NHS there is no significant spare capacity in the system. So how does the system adapt to pressures? Cook et al. (2006)

demonstrated that clinicians modify their behaviour under pressure. This is done dynamically by making goal trade-offs to achieve desirable outcomes, and not by following pre-specified actions. Clinicians have a repertoire of strategies that can be engaged to adapt to pressures (e.g., the physician in charge role: Hosking, Boyle, Ahmed, & Clarkson, 2017; performance repertoire conceptualisation: Furniss, Back, Blandford, Hildebrandt, & Broberg, 2011). However, the use of these opportunistic strategies to manage situations can become exhausted, resulting in decompensation.

In our study of escalation, we found that opportunistic adaptations were used when managing patient flow pressures, and these were more prevalent than the codified actions that were often met with a degraded response (Back et al., 2017). For example, using knowledge of the skill mix of the team allowed one nurse to be engaged in coordinating the setting up of intravenous infusions, allowing less experienced nurses to continue work on other tasks. This was done in response to a codified escalation, where less senior nursing staff who had been sent to help in the department were not qualified to set up infusions.

However, the capacity needed for compensation is often expended; the system becomes brittle, exposing single points of failure and operating in a mode where performance breaches of patient flow targets are accepted as inevitable. In our study, this manifested as 'silo working', where clinicians focused largely on patients assigned to their care but were no longer contributing to patient flow management; this resulted in avoidable throughput target breaches (Back et al., 2017). Informal discussions revealed that this was a conscious strategy of focusing on the one thing they could control – the delivery of care to a current patient. The repertoire of strategies used to maintain an awareness of the system state, and what patient flow activities should be prioritised, was spent.

Are Escalation Policies Purely a Paper-Based Exercise?

The established idea across senior NHS quality improvement managers is that tried and tested escalation policies are needed when managing patient flow pressures (NHS Improvement, 2017). In our study, staff knew that enacting actions specified in the policy would rarely help, and we found that instead escalation could increase the workload at the most inopportune times (Back et al., 2017). Across NHS hospitals, shared anecdotes suggest that staff on the shop floor are reprimanded by management for not adhering to escalation policy (NHS Improvement, 2017). Practitioners suggest that this lack of adherence is because the policies rarely help, and not due to a lack of willingness to change processes to manage pressures. Escalation policies do serve multiple purposes, including satisfying hospital reporting requirements and documenting pressures on the system. This is important, but the policies themselves should also be helpful when managing pressures and should not add to workload because of mandated adherence.

Enhancing Adaptive Capacity across Organisational Boundaries

We argue for conceptual change so that escalation can be better understood and engineered to work more effectively. This is not just a call for redrafted policy, but rather a call for the need to enshrine opportunistic escalation actions as part of the organisation of everyday work, based on in-depth understanding of how this work is achieved. This involves appreciating how to best support the identification, implementation and monitoring of opportunistic escalation actions across organisational boundaries. For example, the use of huddles across a multidisciplinary team, the members of which are not usually collocated, can allow for new pathways to success by sharing pressures and adapting system-wide processes accordingly (*see* Case Study 1). These opportunistic escalation actions are happening all the time as part of 'Work-as-Done' (Anderson, Ross, & Jaye, 2016b; Dekker, 2006). This term refers to how something is actually done, on the 'shop floor', either in a special case or routinely. These actions support the natural capacity in the system for a flexible response to pressures, and should be supported to enhance adaptive capacity. The following case studies highlight escalation actions that we have identified as facilitating resilient performance.

Case Studies

The two case studies presented are based on ethnographic data collected from one large NHS hospital. Over 200 hours of ethnographic data were collected over a period of 30 months. First, we provide a commentary on how patient flow and discharge coordinators (DCos) facilitate the planning of adaptations across multidisciplinary teams. Second, we investigate the reasons why successful escalation requires cross-boundary coordination across the wider hospital system, so that patients are *pulled* rather than being *pushed* around the system.

Case Study 1: Patient Flow and DCos

A watchstander's job on a ship is to maintain awareness so that collisions can be avoided, whilst others are free to get on with their jobs. In hospitals, the watchstanders who ensure patients navigate the system safely are, typically, the nurse in charge (most clinical areas have one), and the most senior physician (who is often only 'on watch' when conducting ward rounds or board rounds). Ideally, the watchstander's role should involve the proactive identification of pressures so that patients can be kept safe. Thus, the watchstander should be involved when identifying the need for escalation. However, often clinical workload prevents these senior members of staff from doing this. Moreover, critical information is often distributed across the multidisciplinary team. The coordination needed to make sense of quickly evolving situations should ideally be supported by the socio-technical design of the system. This is representative of the tension between localised and centralised control, and highlights the need for cross-boundary coordination.

The specialist flow coordinator (FlowCo) role was created in our emergency department to help monitor multiple demands and competing priorities. Every 2 hours, the FlowCo leads a huddle attended by representatives across different areas of the emergency department. The FlowCo presents an overview of pressures; these are discussed by the team. Escalation actions – which are frequently opportunistic rather than codified – are then implemented. On inpatient wards, DCos perform an equivalent role. They help to manage multiple demands and discuss these with the nurse in charge/ward manager as needed. Although these specialist roles are often performed by registered nurses, they are designated non-clinical roles, and civilian clothing is worn to protect against the possibility of being drawn into clinical work.

The FlowCos and DCos act as watchstanders who monitor processes, identify bottlenecks and take actions to improve flow. Some actions, such as negotiating hospital bed availability, chasing missing diagnostic results, finding missing documentation, chasing specialty referrals and locating missing members of staff are focused on solving problems. Identifying the need for other actions, such as adapting organisational processes and resources to pre-empt problems is also a key aspect of the FlowCo and DCo role. This is not straightforward because of the lags in Internet Technology systems and the complexity of patient flows. The FlowCo and DCo help to enhance the capacity for compensation by integrating data from a variety of sources, including IT systems and verbal updates from clinical staff, to make sense of how the system is functioning. They also maintain awareness of the state of the wider hospital system so that possible admission and discharge pathways are identified in anticipation of need, and specialty referrals can be batched to increase efficiency.

The introduction of FlowCo and DCo specialist roles on the shop floor in health care can make the system status more transparent; this allows for opportunistic escalation actions. However, communicating the need for action to others is dependent on how work is organised within a department or ward. In the emergency department, a meeting takes place every two hours and are used to assess the status of patient flow and plan escalation actions. This can only be effective if outcome goals are set and monitored. This requires a member of staff to take ownership over the action, and this is sometimes problematic because of an individual's workload. If the individual does not have the capacity or power to take ownership, we found that the escalation action is less likely to be successful. This leads to a diminishing capacity of the system to adapt to pressures and perform in a resilient way. On the wards, we found that there is an over-reliance on the nurse in charge to organise work by delegation. Therefore, even if opportunistic escalation actions have been identified by the DCo, there might be insufficient task coordination capacity in the system to allow for these actions to be executed.

Case Study 2: Negotiated Pathways – 'Pull' Rather Than 'Push'?

The terms 'push' and 'pull' can be used to describe the type of organisational mechanism used to transfer patients from an emergency department to an appropriate acute inpatient ward. In a 'pull' system, the transition of a patient from the emergency department to an inpatient ward is the responsibility of the inpatient ward. For example, the Acute Frailty Unit (AFU) 'pulls' appropriate patients from the emergency department so that they can be treated by specialist geriatricians. This immediately reduces demand on the emergency department. This is not, however, how patient flow traditionally functions. The emergency department normally assesses and treats patients, and then handles the transfer to the inpatient ward if needed (a 'push' system). Thus, when operating as a 'push' system, the workload is entirely the responsibility of the emergency department. This requires negotiation across boundaries that is rarely acknowledged as part of workflow. Traditionally, when bottlenecks in the system occur, escalation actions aim to ease pressure by streaming or diverting patients (pushing them) to quieter areas of the hospital. Although theoretically sound from a systems control point of view, unintended consequences often emerge that become hard to manage. 'Pull' systems can be created whenever a patient transitions from one point of care to the next and could be more effective than escalation actions that 'push' patients; especially some areas such as the emergency department are often unable to coordinate across boundaries due to pressures of patient numbers.

Prior to the establishment of the AFU, elderly patients were pushed around the hospital system in an attempt to manage patient flow pressures. They could be moved multiple times, creating significant problems for the patients and their carers. This is because assessing, treating and discharging elderly patients is becoming increasingly resource intensive and requires careful coordination across multidisciplinary health care teams within the hospital and in the community. If an elderly patient has been pushed on to an acute assessment ward or elsewhere, there is often insufficient expertise to perform this required coordination effectively. Expertise in the AFU allows for the identification of new conditions amongst a population of patients who are often already known to the hospital and often have comorbidity. This enables expedited decision-making. Effective escalation was found to involve joint working, where a geriatrician attends the emergency department to 'pull' suitable patients to AFU.

The problems associated with a 'push' system became quickly apparent when observing work in the Emergency Medical Unit (EMU). Patients are pushed from the emergency department to the EMU if they require further investigation prior to discharge or admission. During escalation, the unit can be used as an overflow area. If it is determined that a specialist department needs to take over a patient's care, emergency department patients can be temporarily housed in the EMU while they wait to be seen by a specialist. This is problematic because it constrains patient flow and is not the best place for the patient to receive the care that they need. This highlights the fact that although in theory the EMU provides more capacity to assess and treat patients when they arrive, this capacity is quickly eroded by patients awaiting specialty input before being transferred elsewhere. Although escalation actions that are reliant on 'push' systems are less effective, the majority of codified escalation actions involve 'push'. There is a need for joint working across organisational boundaries allowing for opportunistic escalation actions using a 'pull' system. This would potentially increase the system's capacity for compensation.

Discussion and Conclusions

Progress on the journey from the conceptual need for escalation policy, to supporting in situ opportunistic adaptations, is unlikely to be linear. Unfortunately, there is no routine study of what constitutes successful in situ adaptation to patient flow pressures. Instead, it is assumed that the processes as defined in policy are the reason for success or failure. Although individual staff and colleagues may learn from observing others, systematic, organisational-level learning about managing pressures in the system is lacking. Escalation policies throughout the NHS are a manifestation of this lack of understanding.

Successful opportunistic escalation requires the monitoring and anticipation of pressures. Clinicians need to manage patient flow with consideration of resource constraints and variability in care needs. Our ethnographic work found that there are two types of cross-boundary pressures that can be identified and need to be shared to enable opportunistic escalation.

1. *Coordination pressures* – The shared identification of coordination pressures leads to the need to adjust priorities. Being able to share the prerequisites that need to be achieved before key actions can take place can facilitate opportunistic escalation. For example, when expediting discharge, flagging potential delays across the multidisciplinary team can lead to adaptations to processes such as early referrals to the hospital at home team, or warning pharmacy that medication reconciliation is needed in advance.
2. *Misalignment pressures* – Pressures can be discussed in terms of misalignments between demand and capacity. For example, clinicians in

> charge are mandated to report skill mix and staffing misalignments using an acuity and dependency IT tool and escalate as specified in policy. However, the response to this type of escalation is often slow or absent. Sometimes, the time taken to report misalignments would be better spent, sharing them with team members and identifying solutions rather than relying on the IT system. Moreover, a greater awareness of patient flow pressures across a team allows the impact of resource adjustments to be better anticipated. For example, deciding to flex a member of staff elsewhere, to manage a skill mix issue, might cause problems if pertinent information has not been shared with existing team members prior to flexing.

We argue that everyday work should enable cross-boundary coordination across the organisation, ideally across normal work contexts. The cross-boundary coordination refers to work that is specifically designed to reveal connections between micro- and macro-levels in the organisation. In a hospital, cross-boundary coordination can refer to links between small specialist teams, or between ward-level teams and the macro-level, represented by policy, regulations and standards of care. The macro-level is where hospital-wide outcomes are 'performance managed', and assumptions are often made about how processes normally work. For adaptations to be effective, the gap between 'Work-as-Done' and 'Work-as-Imagined' needs to be continually reviewed so that new ways of working can emerge (e.g., Case Study 2 – the AFU pull mechanism), leading to the possibility of multiple paths to successful outcomes.

Cross-boundary mechanisms allow for communication across clinical teams and with managers, regulators and policy makers, and permit the relative merits of opportunistic versus prescribed policy escalation actions to be understood more widely. This could reduce the gulf between 'work-as-imagined' and 'work-as-done'. Such mechanisms will allow us to develop a more comprehensive picture of the pathways to success available to all, and improve our understanding of how best to support the identification, implementation and monitoring of opportunistic escalation actions. This will facilitate resilient performance by enhancing the capacity for effective cross-boundary working.

References

Anderson, J. E., Ross, A. J., Back, J., Duncan, M., Snell, P., Walsh, K., & Jaye, P. (2016a). Implementing Resilience Engineering for Healthcare Quality Improvement Using the CARE Model: A Feasibility Study Protocol. *Pilot and Feasibility Studies*, 2(1), 61.

Anderson, J. E., Ross, A. J., & Jaye, P. (2016b). Modelling Resilience and Researching the Gap Between Work-as-Imagined and Work-as-Done. In J. Braithwaite, R. L. Wears, & E. Hollnagel (Eds.), *Resilient Health Care, Volume 3: Reconciling Work-as-Imagined and Work-as-Done*. Boca Raton, FL: CRC Press.

Back, J., Ross, A. J., Duncan, M. D., Jaye, P., Henderson, K., & Anderson, J. E. (2017). Emergency Department Escalation in Theory and Practice: A Mixed-Methods Study Using a Model of Organizational Resilience. *Annals of Emergency Medicine, 70*(5), 659–671.

Cook, R. & Nemeth, C. (2006). Taking Things in One's Stride: Cognitive Features of Two Resilient Performances. In E. Hollnagel, D. D. Woods, & N. Leveson, (Eds.), *Resilience Engineering: Concepts and Precepts* (pp. 205–220). Aldershot, UK: Ashgate Publishing.

Dekker, S. W. A. (2006). Resilience Engineering: Chronicling the Emergence of Confused Consensus. In E. Hollnagel, D. D. Woods, & N. Leveson (Eds.), *Resilience Engineering: Concepts and Precepts* (pp. 77–92). Boca Raton, FL: CRC Press.

Furniss, D., Back, J., Blandford, A., Hildebrandt, M., & Broberg, H. (2011). A Resilience Markers Framework for Small Teams. *Reliability Engineering & System Safety, 96*(1), 2–10.

Hoffman, R. R. & Woods, D. D. (2011). Beyond Simon's Slice: Five Fundamental Trade-Offs that Bound the Performance of Macrocognitive Work Systems. *IEEE Intelligent Systems, 26*(6), 67–71.

Hollnagel, E., Pariès, J., Woods, D. D., & Wreathall, J. (2011). *Resilience Engineering Perspectives Volume 3: Resilience Engineering in Practice*. Farnham, UK: Ashgate Publishing.

Hosking, I., Boyle, A., Ahmed, V., & Clarkson, J. (2017). What do Emergency Physicians in Charge do? A Qualitative Observational Study. *Emergency Medicine Journal,* 35(3), 186–188.

NHS Improvement. (2017, July). Improving patient flow. Retrieved from https://improvement.nhs.uk/resources/good-practice-guide-focus-on-improving-patient-flow/.

Ombredane, A. & Faverge, J. M. (1955). *L'analyse du travail*. Paris, France: Presses Universitaires de France.

Wears, R. L., Hollnagel, E., & Braithwaite, J. (2015). *Resilient Health Care, Volume 2: The Resilience of Everyday Clinical Work*: Farnham, UK: Ashgate Publishing.

12

Trust and Psychological Safety as Facilitators of Resilient Health Care

Mark A. Sujan, Huayi Huang, and Deborah Biggerstaff
University of Warwick

CONTENTS

Introduction

An important aspect of resilience is the ability to make dynamic trade-offs (Hollnagel, 2009; Sujan, Spurgeon, & Cooke, 2015a). Health care professionals make such dynamic trade-offs in order to deal with mismatches between demand and capacity (Anderson, Ross, & Jaye, 2016), and more generally, to manage gaps, tensions and competing organisational priorities (Cook, Render, & Woods, 2000; Sujan, Rizzo, & Pasquini, 2002, Sujan et al., 2015c). At the interface of care across professional, departmental and organisational boundaries, this requires coordination and negotiation with colleagues and the flexible interpretation of organisational protocols (Nemeth et al., 2007). For the individual clinician, it involves taking interpersonal risks: can I trust my colleague to work towards a shared goal? Do I feel safe in my organisation to bend rules if patient care demands it?

Edmondson, Kramer, & Cook (2004) provide an intuitive characterisation of the concepts of trust and psychological safety in work situations. Trust is defined as an attribute of a relationship with another person. If I trust someone, it means I give them the benefit of the doubt, and I expect that they will behave in certain ways that enable us to achieve our goals in collaboration.

Psychological safety describes individuals' perceptions about the consequences of taking interpersonal risks. It is about whether I believe others will give me the benefit of the doubt.

Previous contributions to the Resilient Health Care (RHC) book series have alluded to the importance of trust and psychological safety. For example, Johnson and Lane (2016) describe the TenC model of resilience, outlining behaviours and characteristics thought to promote RHC. One of the Cs refers to Cohesion, which is defined as mutual respect. Similarly, Braithwaite, Clay-Williams, Hunte and Wears (2016) review a number of studies in emergency care, and conclude that RHC relies on communication, negotiation, teamwork, trust and shared dialogue. Debono and Braithwaite (2015) highlight that a lack of psychological safety can prevent clinicians from talking about the trade-offs they have to make in their everyday clinical work, resulting in a widening of the gap between Work-as-Done by clinicians (WAD) and Work-as-Imagined by managers (WAI).

In this chapter, we continue this line of thinking and aim to connect more explicitly the wider literature on trust and psychological safety to the thinking within RHC. We explore the role of trust and psychological safety in facilitating RHC across boundaries through two examples in which clinicians make trade-offs. The first example is based on a study of patient handover at the boundary between emergency department (ED) and hospital. In this example, we explore how building trust across departmental boundaries allows clinicians to navigate the complexities of delivering cross-disciplinary care. The second example is a typical illustration of the gap between WAI and WAD at the boundary of two professional groups: risk managers and front-line clinicians. Using the specific example of a warfarin discharge policy, we analyse the contribution of psychological safety experienced by front-line clinicians. We argue that psychological safety allows front-line clinicians in this example to make trade-offs with the intention to succeed (i.e., to provide RHC) rather than to prevent blame (by the other professional group: risk managers).

First, we briefly revisit the notion of dynamic trade-offs as a mechanism of RHC.

Dynamic Trade-Offs as a Mechanism of RHC

The delivery of health care is a complex and messy affair. Often, when we think of a system, we have in mind something well designed, linear and clearly bounded. Assembly lines and manufacturing plants come to mind. Health care systems do not fit these traditional conceptions. They are made up of a large number of diverse actors, involving multiple organisations, professions and 'tribes' (Braithwaite, Clay-Williams, Nugus, & Plumb, 2013), often with no clear specification of how individual patients flow through the system. Patients are also not passive recipients of care, but have their own minds, their own diverse histories, personal preferences and current health and social care needs. It could be argued that patients actively co-create their health care through their interactions with health care professionals (Batalden et al., 2016).

Health care professionals need to navigate this complex world supported by organisational protocols and best practices, and their own experience and expertise. Organisational protocols, procedures and guidance represent instances of WAI, i.e., they are assumptions about how clinical work should be done under certain situations. These assumptions usually fall short of capturing the complexity of everyday clinical work due to the large variation in demands and the complexity of the interactions of the different parts in the health care system (Hollnagel, 2016). When we observe WAD, as many of the contributions in this book series have done, we can learn how health care professionals translate gaps and tensions in their everyday clinical work into safe and good quality care by making adaptations and trade-offs. This ability to anticipate and adapt to changes and competing demands by making context-dependent trade-offs is a key mechanism for bringing about RHC (Braithwaite, Wears, & Hollnagel, 2015; Sujan et al., 2015a).

An example described in the second RHC book is the practice of the 'secret second handover' (Sujan et al., 2015b). In this example, paramedics handing over patients to the triage nurse at an ED were faced with competing demands: the need to handover the patient quickly in order to meet demand in the community and the need to communicate in detail the story of the patient under their care. Observations of WAD revealed that paramedics would not always follow the formal protocol designed to meet a time-related performance target. Following the initial handover to the triage nurse, paramedics would sometimes spend additional time in the ED waiting for a nurse to whom they would give a second, thorough handover. This was kept under the radar, because management discouraged the practice as needless duplication. Interviews with paramedics suggested that they used this practice to make dynamic trade-offs (between being in the community quickly vs. giving a thorough handover) on a case-by-case basis, depending on how worried they were about the patient in their care.

The example of the secret second handover illustrates how health care professionals use their experience to make dynamic trade-offs based on what could be regarded a subjective risk assessment of the current situation. However, as we gathered more and more examples about trade-offs from different health care domains, it became clear that these subjective risk assessments were not always acted upon, and that other factors and considerations affect how health care professionals make trade-offs. Two such factors are trust and psychological safety.

Trust – Handing over Patients from the ED to a Hospital Ward

The handover of patients across care boundaries in emergency care was studied in our recent papers (Sujan et al., 2014, 2015d). An example of such a handover is the referral of patients by an ED physician to a hospital specialist. This type of handover is a difficult activity – described as 'selling patients' by ED physicians – because of the perceived need to make patients appear 'attractive' in order to get them admitted onto specialty wards due to bed shortages, staffing levels, increased ward specialisation and more (see also Hilligoss, Mansfield, Patterson, & Moffatt-Bruce, 2015; Nugus et al., 2017; Stephens, Weeds, & Patterson, 2015). The study included discourse analysis of audio recordings of 90 patient referrals from EDs to inpatient wards at three hospitals. The analysis demonstrated that referrals were not simply unidirectional conversations, where the sender (the ED physician) handed over information to a more or less passive receiver (ward clinician). Instead, the conversations entailed both descriptive as well as collaborative phases, and leadership and initiative could switch between the communication parties.

In practice, both parties are involved in trade-offs: ED physicians need to find a balance between moving the patient quickly to avoid ED overcrowding and breaching time-related performance targets and providing a sufficiently robust diagnosis that enables acceptance onto a ward, but which might require additional time and effort. On the other hand, clinicians on wards need to balance the recognised need of the ED for fast patient disposition, with considerations of their own workload and concerns that the patient goes to the correct, or most suitable, place first time. During the collaborative phase of the referral conversation, the ED physician and the clinician on the ward would aim to reach a shared understanding of the patient's situation and needs and try to arrive at a joint decision about patient disposition.

The discourse analysis illustrated that sometimes the collaborative phase of the referral conversation supported resilience through effective coordination and negotiation, whereas in other situations, the conversation was difficult and frustrating for both parties. We believe that the level (or lack) of trust between the two parties, as well as the strength of their other trust relationships (e.g., other colleagues, managers, organisational norms and protocols) might help explain, in part, the dynamics of the referral conversation. Trust has been framed in different ways, but in general, it relates to overcoming a sense of perceived vulnerability over time that originates from uncertainty about the motives and future actions of the other person (Kramer, 1999); it involves our ability to rely on another person's behaviour or actions in a situation where there is a perceived risk of the other party taking advantage of us (Williams, 2001). Studies have shown that trust can have a number of benefits, as it can function as a social heuristic, thereby reducing transaction costs, and can trigger positive cooperative forms of behaviour (Uzzi, 1997). Comparatively, distrust makes people more suspicious of the other person's intentions, which can prompt more active and careful consideration of what the other person is doing and perhaps thinking (Fein, 1996). Trust can arise in different ways; for example, through disposition (i.e., being a trusting person), from the previous history of a relationship, or based on perceptions of roles (e.g., being a doctor makes one appear trustworthy in the eyes of others) (Kramer, 1999). It has been shown that trust is fragile and easier to destroy than to build (Meyerson, Weick, & Kramer, 1996).

In the referral conversation, the ED physician and the hospital clinician experience uncertainty about each other's motives and intentions. This uncertainty may be heightened due to the structural characteristics of ED referrals. Research suggests that trust is particularly relevant for team performance in situations where there is high task interdependency, where team members cannot interact face-to-face, where team membership is changing frequently, and where there are strong differentiations in authority and specialty (de Jong, Dirks, & Gillespie, 2016). These characteristics are typically present in ED referrals, which take place on the phone, between an ED generalist and a hospital specialist of different ranks, who might not have interacted before. In a situation of such uncertainty, a trusting relationship enables them to give each other the benefit of the doubt and assume that they are both working in the patient's best interest. ED physicians need to overcome suspicions that hospital clinicians are gatekeeping and creating fences around their wards, and that they only have an interest in their own workflows. Ward clinicians need to set aside doubts about the appropriateness of referrals, and suspicions that omissions have been forced by using specific keywords or omitting details (e.g., not mentioning surgical history when referring to a medical ward).

Trust and distrust can be based on a combination of different sources. Some clinicians might have collaborated on previous occasions with positive outcomes and have established trust over time. Trust can arise from an expectation of expertise that is associated with a particular role, such as a consultant. However, many clinicians have previous negative experiences, either with their current communication partner or with other people in similar professional situations; this can undermine trust quickly and threaten their willingness for cooperation. There can also be an element of role-based distrust, where more senior doctors in the ED assume that junior doctors on the wards are not completely familiar with referral pathways, and where senior clinicians on the ward doubt the ability of juniors in the ED to properly work up a robust diagnosis of complex patient presentations.

The clinicians involved in the referral conversation form a number of trust relationships in addition to the trust relationship between themselves. Trust and distrust can be based on organisational procedures and norms. For example, ED physicians might distrust the referral pathways, perceiving them as unfair and lacking a 'pulling' mechanism (i.e., wards are waiting for the ED to 'push' patients towards them rather than seeking out patients proactively). Ward clinicians might similarly distrust the referral system, perceiving the ED as creating (inappropriate) work for wards. ED physicians and ward clinicians will also have established relationships of trust or distrust with senior decision makers in the hospital (e.g., medical director), and these might be called upon to mediate in some cases. In conclusion, these multiple trust relationships, and the different foundations of trust for each trust relationship, can influence the referral conversation and impact on the outcome.

Study participants provided a number of practical suggestions for how the referrals from the ED to hospital wards might be improved. The mechanisms of these suggestions can be understood from a trust perspective. Participants emphasised the need for joint working and the adoption of a systems approach. Trust within a heterogeneous group of health care professionals from different disciplinary and cultural backgrounds might be enhanced by developing and promoting personal relationships (being able to 'put a face to a name') and by establishing mutual awareness of each other's goals and motivations. A practical solution put forward by participants was the idea of shadowing, where health care professionals from one domain would have the opportunity to observe work in another domain to build awareness and reduce suspicion and safeguarding behaviours. Joint appointments were also suggested, where an individual would be employed by different organisations (e.g., the ambulance service and the ED) in order to bridge the cultural divide. This is a way of modifying the social structure of organisations, with the intention of building trust across organisational boundaries (McEvily, Perrone, & Zaheer, 2003).

Psychological Safety – Discharging Patients Who Are on Warfarin

Sujan, Huang, & Braithwaite (2016a) described an example where an inner-city hospital had experienced a number of incidents involving patients who had been discharged with prescriptions for warfarin (a blood thinner). Such incidents with warfarin include bleeding following a fall at home due to elevated international normalised ratio (INR) and the occurrence of thrombo-embolism due to low INR. Warfarin can interact with other medication and can affect INR levels in different ways for different patients. Hence, regular INR measurements are vital to ensure appropriate warfarin management. For this patient group, it is very important to ensure that their drug regime is appropriate and that they follow this regime once they are back at home – otherwise, they run the risk of over- or under-thinning their blood. Some patients find this very difficult, and a checkup is advised to review their progress.

The hospital conducted standard root cause analyses (RCA) of the adverse events. The RCA's resulted in a recommendation for the introduction of a new discharge policy. This policy specified that patients who are on warfarin could only be discharged with a follow-on appointment for the anti-coagulation clinic in place. On face value, this policy sounded eminently sensible and safe practice. However, the appointment booking service was only available only during weekday office hours. For clinicians, this created a tension and difficult decisions. If they had a patient who could be discharged on a Friday evening or Saturday morning, should they follow the policy and keep the patient in hospital without clinical need until the appointment could be booked the following Monday? Or would it be better to discharge the patient without a follow-on appointment in place? Either solution is problematic. Keeping patients in hospital longer than necessary subjects them to risks, such as hospital-acquired infections, and wastes crucial hospital resources. Violating the policy and sending patients home without follow-on appointments might result in similar adverse events. How do clinicians make trade-offs in these situations to resolve the tension?

On the one hand, clinicians appear to make a subjective risk assessment to determine the risk of discharge without follow-on appointments for individual patients. If a patient is lucid, sensible and mobile, they might be discharged with a request to return to the ward in person to ensure that they have a follow-on arranged for them. On the other hand, not every clinician will opt to violate the policy for low-risk patients. A potential explanation for this is that an incident report will be triggered and recorded automatically when a patient comes back to book their follow-on appointment. This might have repercussions for the clinician who discharged the patient.

Violating the policy clearly carries risk for clinicians. We hypothesise that their level of psychological safety influences decisions about whether to discharge a patient. Psychological safety describes the degree to which people perceive their work environment as being supportive of interpersonally risky behaviours (Edmondson, Higgins, Singer, & Weiner, 2016), such as bending rules or reporting problems. Psychological safety can support teams and organisations in learning behaviours and improved performance, because staff are less likely to focus on self-protection. Instead, staff can engage more freely in information sharing, exploration of divergent ideas, risk taking and exploratory learning (Edmondson & Lei, 2014). Research on incident reporting and organisational learning in health care suggests that in environments that fail to provide psychological safety, or in blame cultures, staff are less likely to speak up about problems they experience in everyday clinical work (Sujan, 2015; Tucker & Edmondson, 2003). Psychological safety can be improved by leadership behaviours, such as reducing the hierarchical gap in authority between roles (e.g., between doctors and nurses), actively seeking and valuing staff input, and acknowledging fallibility (Edmondson et al., 2016).

In the previous example, clinicians may not feel sufficiently safe to adapt the discharge policy using their subjective risk assessment and experience if negative precedents have been set, and if the organisation is perceived to blame people for violations and mistakes. However, if the organisational leadership actively encourages reporting of problems, clinicians might resolve the trade-off in this situation according to their assessment of what is best for the patient and the hospital. Clinicians and management can engage to identify solutions for reducing the gap between WAI and WAD that was opened up by the introduction of the discharge policy.

Discussion and Conclusion

The two examples described earlier illustrate how the concepts of trust and psychological safety might provide additional insights into how people make dynamic trade-offs, and thereby further our understanding of RHC across boundaries. The first example considers how clinicians make trade-offs at and across the boundary of the ED and the hospital. The second example shows how front-line clinicians make trade-offs within a context set by another professional group (i.e., procedures and policies that represent WAI by risk manager).

Trust can influence whether, and with whom, people seek to engage and negotiate, and how they approach this (McEvily et al., 2003). If there is trust between the two parties, we expect to see more open exchanges and joint problem solving, and less defensive behaviours, because people don't need to worry as much about protecting themselves against potential harm from others (Mayer & Gavin, 2005). This should result in better performance (de Jong et al., 2016), and, by extension, more RHC. Psychological safety can support people in adapting rules and policies and in speaking up about this. We believe that this might encourage people to make trade-offs based on their own assessment of the requirements of a specific situation and less based on static protocols and organisational targets. Psychological safety also contributes to increased organisational learning, because people can report and share the adaptations they need to make. This should reduce the gap between WAI and WAD and prompt further improvement activities.

Arguably, health care as a domain has a poor track record of recognising and incorporating such non-technical considerations into educational and improvement efforts. Health care is still largely based on deeply ingrained notions of clinical autonomy, hierarchical structures and professional identities. This can prevent people at lower levels of the hierarchy from speaking up about problems, and it can hinder the flow of information across boundaries (Edmondson et al., 2016). In addition, most of the structural characteristics that make trust between health care professionals particularly important are present at the interface of care: task interdependency, team virtuality, temporal team membership, and authority and skill differentiation (de Jong et al., 2016). Yet, in practice, it is in precisely those situations where trust is needed most that establishing, maintaining and rebuilding trust is most difficult.

What are the implications for RHC? Arguably, a first step is to raise awareness and share knowledge of the importance of the concepts of trust and psychological safety among health care professionals, managers and RHC researchers. The appreciation of trust might prompt organisations to focus more on initiatives that provide conditions in which trust can arise, such as investment in communities of practice, and informal learning and improvement activities (Sujan, 2015). Health care professionals frequently interact with individuals with whom they have never collaborated before. In these situations, the foundation for trust might involve the perceived trustworthiness of a given role (e.g., 'the orthopaedic surgeon'), rather than that of a particular individual. It might be expected that resilience will be weakened in situations where role-based trust is poor, whereas it might be strengthened where a role is regarded as more trustworthy. Research suggests that it is at the boundaries of organisations that role autonomy elicits greater levels of trust, allowing interacting parties to respond to each other's needs more effectively (Perrone, Zaheer, & McEvily, 2003).

Within the RHC community, a number of interesting avenues to support organisational learning by encouraging staff to speak up have already been proposed, such as the combined Safety-I and Safety-II approach to improvement described in Chuang & Wears (2015), learning from the ordinary (Sujan, 2012; Sujan, Pozzi, & Valbonesi, 2016c) and learning from excellence (Kelly, Blake, & Plunkett, 2016). Common to these approaches is the recognition that useful learning can be derived from the gap between WAI and WAD, rather than focusing only on adverse events and outcomes (which typically leads to efforts of increasing managerial control by bringing WAD closer to WAI) (Sujan et al., 2016c). Edmondson et al. (2016) also suggest that psychological safety should be a factor already considered in the curricula of student health care professionals.

By integrating the existing body of knowledge on trust and psychological safety, the field of RHC can develop strategies to promote trusting relationships and establish work environments in which people feel safe to take interpersonal risks.

Acknowledgement

We are grateful to health care professionals participating in the Warwick Medical School Masters module 'Improving safety and quality of healthcare' for sharing their experiences of making trade-offs in everyday clinical work.

References

Anderson, J. E., Ross, A. J., & Jaye, P. (2016). Modelling Resilience and Researching the Gap Between Work-as-Imagined and Work-as-Done. In J. Braithwaite, R. Wears, & E. Hollnagel (Eds.), *Resilient Health Care III: Reconciling Work-as-Imagined with Work-as-Done* (pp. 133–152). Farnham, UK: Ashgate Publishing.

Batalden, M., Batalden, P., Margolis, P., Seid, M., Armstrong, G., Opipariarrigan, L., & Hartung, H. (2016). Coproduction of Healthcare Service. *BMJ Quality & Safety, 25*, 509–517.

Braithwaite, J., Clay-Williams, R., Hunte, G., & Wears, R. (2016). Understanding Resilient Clinical Practices in Emergency Department Ecosystems. In J. Braithwaite, R. Wears, & E. Hollnagel (Eds.), *Resilient Health Care, Volume 3: Reconciling Work-as-Imagined and Work-as-Done* (pp. 89–102). Farnham, UK: Ashgate Publishing.

Braithwaite, J., Clay-Williams, R., Nugus, P., & Plumb, J. (2013). Healthcare as a Complex Adaptive System. In E. Hollnagel, J. Braithwaite, & R. Wears (Eds.), *Resilient Health Care* (pp. 57–71). Farnham, UK: Ashgate Publishing.

Braithwaite, J., Wears, R. L., & Hollnagel, E. (2015). Resilient Health Care: Turning Patient Safety on Its Head. *International Journal for Quality in Health Care, 27*(5), 418–420.

Chuang, S. & Wears, R. (2015). Strategies to Get Resilience into Everyday Clinical Work. In R. Wears, E. Hollnagel, & J. Braithwaite (Eds.), *The Resilience of Everyday Clinical Work* (pp. 225–234). Farnham, UK: Ashgate Publishing.

Cook, R. I., Render, M., & Woods, D. D. (2000). Gaps in the Continuity of Care and Progress on Patient Safety. *BMJ, 320*(7237), 791–794.

Debono, D. & Braithwaite, J. (2015). Workarounds in Nursing Practice in Acute Care: A Case of a Health Care Arms Race? In R. Wears, E. Hollnagel, & J. Braithwaite (Eds.), *The Resilience of Everyday Clinical Work* (pp. 23–38). Farnham, UK: Ashgate Publishing.

De Jong, B. A., Dirks, K. T., & Gillespie, N. (2016). Trust and Team Performance: A Meta-Analysis of Main Effects, Moderators, and Covariates. *Journal of Applied Psychology, 101*(8), 1134–1150.

Edmondson, A. C., Higgins, M., Singer, S., & Weiner, J. (2016). Understanding Psychological Safety in Health Care and Education Organizations: A Comparative Perspective. *Research in Human Development, 13*(1), 65–83.

Edmondson, A. C., Kramer, R. M., & Cook, K. S. (2004). Psychological Safety, Trust, and Learning in Organizations: A Group-Level Lens. In K. S. Cook & R. M. Kramer (Eds.), *Trust and Distrust in Organizations: Dilemmas and Approaches* (pp. 239–271). New York, NY: Russell Sage Foundation.

Edmondson, A. C. & Lei, Z. (2014). Psychological Safety: The History, Renaissance, and Future of an Interpersonal Construct. *Annual Review of Organizational Psychology and Organizational Behavior, 1*(1), 23–43.

Fein, S. (1996). Effects of Suspicion on Attributional Thinking and the Correspondence Bias. *Journal of Personality and Social Psychology, 70*(6), 1164–1184.

Hilligoss, B., Mansfield, J. A., Patterson, E. S., & Moffatt-Bruce, S. D. (2015). Collaborating—or "Selling" Patients? A Conceptual Framework for Emergency Department-to-Inpatient Handoff Negotiations. *Joint Commission Journal on Quality and Patient Safety, 41*(3), 134–143.

Hollnagel, E. (2009). *The ETTO Principle: Efficiency-Thoroughness Trade-Off*, Farnham, UK: Ashgate Publishing.

Hollnagel, E. (2016). Prologue: Why do Our Expectations of How Work Should be Done Never Correspond Exactly to How Work is Done. In J. Braithwaite, R. Wears, & E. Hollnagel (Eds.), *Resilient Health Care Volume 3: Reconciling Work-as-Imagined and Work-as-Done* (pp. xvii–xxv). Farnham, UK: Ashgate Publishing.

Johnson, A. & Lane, P. (2016). Resilience Work-as-Done in Everyday Clinical Work. In J. Braithwaite, R. Wears, & E. Hollnagel (Eds.), *Resilient Health Care Volume 3: Reconciling Work-as-Imagined with Work-as-Done* (pp. 71–88). Farnham, UK: Ashgate Publishing.

Kelly, N., Blake, S., & Plunkett, A. (2016). Learning from Excellence in Healthcare: A New Approach to Incident Reporting. *Archives of Disease in Childhood, 101*(9), 788–791.

Kramer, R. M. (1999). Trust and Distrust in Organizations: Emerging Perspectives, Enduring Questions. *Annual Review of Psychology, 50*(1), 569–598.

Mayer, R. C. & Gavin, M. B. (2005). Trust in Management and Performance: Who Minds the Shop While the Employees Watch the Boss? *Academy of Management Journal, 48*(5), 874–888.

McEvily, B., Perrone, V., & Zaheer, A. (2003). Trust as an Organizing Principle. *Organization Science, 14*(1), 91–103.

Meyerson, D., Weick, K. E., & Kramer, R. M. (1996). Swift Trust and Temporary Groups. In R. M. Kramer & T. R. Tyler (Eds.), *Trust in Organizations: Frontiers of Theory and Research* (pp. 166–195). Thousand Oaks, CA: Sage Publications.

Nemeth, C. P., Nunnally, M., O'Connor, M. F., Brandwijk, M., Kowalsky, J., & Cook, R. I. (2007). Regularly Irregular: How Groups Reconcile Cross-Cutting Agendas and Demand in Healthcare. *Cognition, Technology and Work, 9*(3), 139–148.

Nugus, P., McCarthy, S., Holdgate, A., Braithwaite, J., Schoenmakers, A., & Wagner, C. (2017). Packaging Patients and Handing Them Over: Communication Context and Persuasion in the Emergency Department. *Annals of Emergency Medicine, 69*(2), 210–217.

Perrone, V., Zaheer, A., & McEvily, B. (2003). Free to be Trusted? Organizational Constraints on Trust in Boundary Spanners. *Organization Science, 14*(4), 422–439.

Stephens, R., Weeds, D., & Patterson, E. (2015). Patient Boarding in the Emergency Department as a Symptom of Complexity-Induced Risks. In R. Wears, E. Hollnagel, & J. Braithwaite (Eds.), *Resilient Health Care, Volume 2: The Resilience of Everyday Clinical Work* (pp. 159–174). Farnham, UK: Ashgate Publishing.

Sujan, M. A. (2012). A Novel Tool for Organisational Learning and its Impact on Safety Culture in a Hospital Dispensary. *Reliability Engineering & System Safety, 101*, 21–34.

Sujan, M. A. (2015). An Organisation Without a Memory: A Qualitative Study of Hospital Staff Perceptions on Reporting and Organisational Learning for Patient Safety. *Reliability Engineering & System Safety, 144*, 45–52.

Sujan, M. A., Huang, H., & Braithwaite, J. (2016a). Why do Healthcare Organisations Struggle to Learn from Experience? A Safety-II Perspective. In *The 2016 Healthcare Systems Ergonomics and Patient Safety Conference (HEPS 2016)*, Toulouse, France.

Sujan, M. A., Huang, H., & Braithwaite, J. (2016b). Learning from Incidents in Health Care: Critique from a Safety-II Perspective. *Safety Science, 99*, 115–121.

Sujan, M. A., Pozzi, S., & Valbonesi, C. (2016c). Reporting and Learning: From Extraordinary to Ordinary. In J. Braithwaite, R. Wears, & E. Hollnagel (Eds.), *Resilient Health Care, Volume 3: Reconciling Work-as-Imagined with Work-as-Done* (pp. 103–110). Farnham, UK: Ashgate Publishing.

Sujan, M. A., Rizzo, A., & Pasquini, A. (2002). Contradictions and Critical Issues during System Evolution. In *ACM Symposium on Applied Computing*, Madrid.

Sujan, M. A., Spurgeon, P., & Cooke, M. (2015a). The Role of Dynamic Trade-Offs in Creating Safety—A Qualitative Study of Handover Across Care Boundaries in Emergency Care. *Reliability Engineering & System Safety, 141*, 54–62.

Sujan, M. A., Spurgeon, P., & Cooke, M. (2015b). Translating Tensions into Safe Practices Through Dynamic Trade-Offs: The Secret Second Handover. In R. Wears, E. Hollnagel, & J. Braithwaite (Eds.), *The Resilience of Everyday Clinical Work* (pp. 11–22). Farnham, UK: Ashgate Publishing.

Sujan, M. A., Chessum, P., Rudd, M., Fitton, L., Inada-Kim, M., Cooke, M. W., & Spurgeon, P. (2015c). Managing Competing Organizational Priorities in Clinical Handover Across Organizational Boundaries. *Journal of Health Services Research & Policy, 20*, 17–25.

Sujan, M. A., Chessum, P., Rudd, M., Fitton, L., Inada-Kim, M., Spurgeon, P., & Cooke, M. W. (2015d). Emergency Care Handover (ECHO Study) Across Care Boundaries: The Need for Joint Decision Making and Consideration of Psychosocial History. *Emergency Medicine Journal, 32*(2), 112–118.

Sujan, M. A., Spurgeon, P., Inada-Kim, M., Rudd, M., Fitton, L., Hornblow, S., Cross, S., Chessum, P., & Cooke, M. (2014). Clinical Handover within the Emergency Care Pathway and the Potential Risks of Clinical Handover Failure (ECHO): Primary Research. *Health Services Delivery Research, 2*(5).doi: 10.3310/hsdr02050.

Tucker, A. L. & Edmondson, A. C. (2003). Why Hospitals Don't Learn from Failures: Organizational and Psychological Dynamics that Inhibit System Change. *California Management Review, 45*(2), 55–72.

Uzzi, B. (1997). Social Structure and Competition in Interfirm Networks: The Paradox of Embeddedness. *Administrative Science Quarterly, 42*, 35–67.

Williams, M. (2001). In Whom We Trust: Group Membership as an Affective Context for Trust Development. *The Academy of Management Review, 26*(3), 377–396.

13

Collaborative Use of Slack Resources as a Support to Resilience: Study of a Maternity Ward

Natália Basso Werle, Tarcisio Abreu Saurin, and Marlon Soliman

Federal University of Rio Grande do Sul

CONTENTS

Introduction

In complex socio-technical systems such as health care, performance variability is part of everyday work and a source of both desired and undesired outcomes (Hollnagel, 2012). As a support for coping with variability, slack is also part of everyday work in health care. Slack is a mechanism for reducing interdependencies and minimising the possibility of one process affecting another, and thus it makes processes loosely coupled (Safayeni & Purdy, 1991). In a similar definition, slack can be framed as a cushion of current or potential resources that enable an organisation to successfully adapt to internal or external pressures (Bourgeois, 1981), threats and opportunities. Thus, the use of slack does not necessarily mean the system's functionality is impaired (Saurin & Werle, 2017). In fact, slack is usually operationalised through some type of human (e.g., cross-trained professionals), technical (e.g., spare pieces of equipment) or organisational resource (e.g., double-check of medical prescriptions), which make it an integral part of health care.

In this chapter, we focus on situations in which several slack resources are jointly deployed, supporting collaborative work across professional, departmental and institutional boundaries; hereafter, the term *collaborative work* is used in this sense of working across the said boundaries. Collaborative work is characterised by team problem-solving and sharing responsibilities for achieving common goals (Schöttle, Haghsheno, & Gehbauer, 2014), and is key for health care resilience and safety (Greenhalgh, 2008). Wachs, Saurin, Righi, and Wears (2016) identified collaborative work as the most frequent category of resilience skill used by professionals in two emergency departments in Brazil and the United States. Bardram and Bossen (2005) discussed how non-digital artefacts, such as whiteboards, work schedules and case records, supported collaborative work in hospital wards. The need for redundant information within the system of artefacts was also analysed by those authors – for example, information about a patient's room was repeated on the work schedule and whiteboards. Ong and Coiera (2010) reported the use of redundant procedures for preventing errors during inpatient transfers to radiology, which was a process that involved collaboration between several professionals.

Collaboration in cognitively demanding settings necessarily involves at least one form of slack, namely cognitive diversity, which refers to a divergence in analytical perspectives among members of an organisation (Schulman, 1993). The modelling of collaborative work where a broader range of slack resources is used can shed light on their complementary roles as well as on possibilities for redesigning their interactions. Also, this modelling can produce useful data (e.g., examples of adverse events prevented by a pool of slack resources) to defend slack against efficiency pressures. This chapter discusses the collaborative use of slack across boundaries in a maternity ward. This discussion is based on data originally collected by Saurin and Werle (2017) in the context of a study for developing a framework for the analysis of slack in socio-technical systems.

Classification of Slack

Saurin and Werle (2017) proposed a classification of slack, which is summarised as follows:

1. **Origin:** slack may either be designed, which usually occurs in tightly coupled systems, or opportunistic, which usually occurs in loosely

coupled systems, in which slack is often intrinsic to their nature (Perrow, 1984). Designed slack arises from proactive organisational resilience, while opportunistic slack relies on reactive individual and team resilience.

2. **Nature of the resources:** in principle, any physical or virtual resource can work as slack in a certain context, although the resources typically considered are time, people, materials, space and money. There may be resources that are more elusive and difficult to be quantified, such as perspectives to solve a problem and degrees of freedom in standard operating procedures.
3. **Availability:** slack may either be immediately available or not. Availability is easier if slack is near the point of use, decentralised, and users have autonomy to deploy the resources (Sharfman, Wolf, Chase, & Tansik, 1998). Furthermore, availability depends on whether slack is protected from undesired use by agents either internal or external to the system. For instance, some intensive care units have a spare bed for urgent incoming patients, and this bed should not be occupied by patients who do not need critical care (Silich et al., 2011). One means of supporting this protection is through real-time visibility of the status of slack. In order to support the availability assessment, in the context of front-line health care operations, we propose three arbitrarily defined criteria as follows: (i) high availability, if the slack resource can be deployed in less than 5 min; (ii) moderate availability, between 5 and less than 30 mins; and (iii) low availability, if more than 30 mins.
4. **Visibility:** the availability of extant slack should be easily and quickly visible in the workplace, to support performance adjustments triggered by scarcity of resources. As a basis for evaluating the visibility of slack in front-line health care operations, three criteria are proposed: (i) high visibility, when the status of slack is visible in real-time (e.g., through boards, publicly displayed screens or direct observation in the surrounding work environment) without the need for verbal communication or checking information in the computerised system; (ii) moderate visibility, when there is a need for verbal exchange of information and/or checking the computerised system and (iii) low visibility, when the conditions defined for high and moderate visibility do not hold, or when it is not possible to know the status of the slack resources.
5. **Strategy of deployment:** five strategies were identified. The first strategy, redundancy, can be divided into several sub-categories, such as standby redundancy, active redundancy and duplication of functions (Clarke, 2005). This author provides definitions focused

on human redundancy, although the strategy is also applicable to other resources. For instance, standby human redundancy implies the redundant individual is not immediately involved in the task at hand, is typically not present in the operator's immediate environment and must be called when necessary (Clarke, 2005). Active redundancy means the individual performing a redundant function is involved in the task at hand – e.g., a worker carries out a task while another monitors the performance of that operator (Clarke, 2005).

The second strategy for the deployment of slack is through the design of work-in-process, which refers to the creation of queues between workstations. This strategy is widely used in manufacturing plants, and the amount of work-in-process is a function of the stability of processes; the more unstable, the greater the work-in-process (Liker, 2004).

The third strategy refers to three types of margins of manoeuvre suggested by Stephens, Woods, Branlat, and Wears (2011). Margin of manoeuvre type 1 is characterised by maintaining local margin by restricting other units' actions or borrowing other units' margin. Margin type 2 accounts for autonomous strategies to create margin via local reorganisation or expand a unit's ability to regulate its margin. Type 3 refers to coordinated, collective action of recognising or creating a common-pool resource on which two or more units can draw (Stephens et al., 2011).

The fourth strategy is conceptual slack or cognitive diversity, which refers to a divergence in analytical perspectives among members of an organisation. The fifth strategy is control slack, which implies individual degrees of freedom in organisational activity, with some range of individual action unconstrained by formal structures of coordination or command. The fourth and fifth strategies were proposed by Schulman (1993).

6. **Side effects:** in complex socio-technical systems, elements are highly interconnected and recursively influence each other. Therefore, the introduction of slack is not a neutral action, making it necessary to assess side effects, such as new possibilities of error and maintenance costs.
7. **Durability:** this category refers to how long slack maintains its properties, even if it is not deployed, and to how often it needs to be replaced. The rate of degradation can be non-linear – e.g., technological or organisational changes can simply render a given type of slack irrelevant to its original intention.
8. **Scope:** this refers to the breadth of sources of variability that a slack resource can match. The more sources of variability can be matched

the more general purpose the slack is. Scope seems to be related to the nature of resources, since some of them are intrinsically more general purpose (e.g., money). An important dimension of scope is related to the adaptability of slack, which is associated with the idea that slack can self-adjust to dynamic variability. In order to support the assessment of scope in this study, we propose that (i) wide scope slack resources cover, at least partially, more than 60% of known variability sources; (ii) moderate scope resources cover from 30% to 60% of variability sources and (iii) low scope resources cover less than 30% of variability sources.

9. **Legal requirement:** several slack resources are demanded by mandatory regulations, which may include technical and management specifications related to the previously mentioned categories.

Table 13.1 presents the implications of the aforementioned classification for collaborative work. In order to support collaborative work, slack should ideally bear specific characteristics, such as immediate availability and high visibility.

TABLE 13.1

Classification of Slack and Its Implications for Collaborative Work

Category	Implications for Collaborative Work
Origin	Designed slack tends to be more useful for collaborative work, since it may be more visible and accessible by all interested parties
Nature of the resources	Slack resources formed by people tend to provide direct support for collaborative work. Other resources (e.g., equipment) can play a role as moderators of interactions between people
Availability	Slack that is immediately available and located near to the point of use tends to be more useful for collaborative work
Strategy of deployment	Some deployment strategies by definition involve collaborative work – e.g., standby and active human redundancy, margin of manoeuvre type 3 and cognitive diversity
Visibility	Highly visible slack tends to be more useful for collaborative work, since this helps team members to develop a shared mental model of the system status. In turn, low-visibility slack can encourage 'wasteful' collaborative work, since employees may need to ask colleagues about the location and status of the slack
Side effects	Too much slack can discourage collaborative work: processes can become loosely coupled and isolated from each other
Durability	No clear implication
Scope	No clear implication
Legal requirement	Regulations that mandate the use of some forms of slack may include requirements of collaborative work – e.g., joint decision-making

Empirical Study

Overview of the Studied Maternity Ward

The maternity ward was part of a 380-bed private hospital in Southern Brazil, which is recognised by the Ministry of Health as a leading institution in terms of quality of care. The maternity ward is composed of 32 inpatient beds, four operating rooms, a recovery room (seven beds), an obstetric emergency department open 24/7 and a neonatal Intensive Care Unit (ICU) (27 beds). These facilities were spread over three floors and allowed up to 17 births a day. According to data provided by the hospital management, about 80% of births occurred through caesarean section, and the remaining 20% were vaginal births. This rate of caesarean section is much higher than the 10%–15% recommended by the World Health Organization, although it is not unusual in Brazilian hospitals (Betrán et al., 2007). While caesarean sections allow for anticipated scheduling of births, they also imply a longer length of stay for patients: up to 72 hours, in contrast to 48 hours for vaginal births.

Obstetricians and other physicians who assist the deliveries are service providers to the hospital (390 in total) rather than employees. As such, these physicians are seen by hospital managers as 'clients', since in principle the deliveries of their private patients could be scheduled to other hospitals. A portion of the clinical staff is made up of hospital employees, who work mostly in the obstetric emergency department. The maternity ward's own staff includes 87 professionals: 12 obstetricians, 17 nurses and 58 nurse technicians.

Analysing the Collaborative Use of Slack in the Maternity Ward

The collaborative use of slack was analysed based on the discussion of two incidents requiring the provision of emergency care in the maternity ward. These events were identified using interviews with professionals through the critical decisions method. Each event was modelled through the Functional Resonance Analysis Method (FRAM) (Hollnagel, 2012). The traditional steps for applying the FRAM were followed by (i) identifying and describing the functions that played a role in the event, according to the six aspects of each function, namely input, output, precondition, resource, control and time; (ii) analysing the variability of the output of each function, both in terms of time and precision and (iii) identifying couplings between functions.

Considering the objectives of this study, the analysis of FRAM models emphasised: (i) the identification of the functions whose outputs had significant variability and therefore triggered the need for slack; (ii) the identification of functions that deployed the use of slack and (iii) the analysis of how variability and slack crossed professional, departmental and institutional boundaries. For each model, overall scores of the risks associated with variability sources and the effectiveness of slack were estimated and compared, based on questionnaires answered by the professionals who worked in the maternity ward.

Saurin and Werle (2017) describe the data collection and analysis procedures for obtaining the scores in detail, and thus only an overview is presented in this section. Regarding the identification of slack, the concept of Unit of Slack was a starting point. Each pool of similar slack resources that shared the same purpose was defined as a Unit of Slack. The Units of Slack were identified based on multiple sources of data, such as critical decisions method interviews with seven professionals, 25 hours of non-participant observation, 6 hours of participant observations and analysis of documents. A content analysis of the transcripts of interviews and notes from observations was made in order to spot excerpts of text that matched the previously mentioned definition of slack.

An analysis of the effectiveness of the Units of Slack was then conducted. Professionals answered a survey with the following question: how effective is this slack? The questionnaires had a 15 cm scale and two anchors at the extremes: low effectiveness and high effectiveness. Average effectiveness scores for each Unit of Slack were calculated, and these were then transformed into a 100-point scale.

Variability sources were also identified through a content analysis from the same database produced for spotting slack. Next, a risk index associated with each variability source was calculated. Risk was the result of multiplying scores related to the frequency (how often it appears) and severity (how strong its impact is) of each variability source. All variability sources were listed in two questionnaires: one focused on the assessment of frequency and the other focused on severity. In each questionnaire, respondents had to make a mark on a 15 cm scale, indicating how much they thought the sources of variability were frequent and impacting on safety and efficiency. Thus, the anchors of the questionnaires were as follows: rarely and very often (for assessing frequency); and low impact and high impact (for assessing severity). Average results of risk were transformed into a 100-point scale. Forty-five of the 87 professionals in the ward voluntarily answered the questionnaires. On average, respondents had 7.5 years of experience in health care, and their distribution across professional categories was as follows: technicians (71%), nurses (18%) and physicians (11%).

Results

Event 1

Event 1 was reported by one of the interviewed obstetricians. According to her report, a pregnant woman arrived at the obstetrics emergency unit late at night, complaining that she was feeling unwell. After some exams, physicians determined that the foetus was dead. At this moment, the first slack resource was deployed by calling on a psychologist on duty to comfort the patient. Physicians then decided to induce an abortion and kept the patient under observation. However, during this process the patient began to haemorrhage and required emergency surgery.

As such, there were two functions whose outputs triggered the need for slack (Figure 13.1): <examine patient>, <induct abortion>. In turn, the five triggered slack functions were: <call psychologists on duty>, <call obstetrician on duty>, <call quick response team>, <use of team expertise>, <call external surgeon and anaesthetist>. The outputs of these slack functions provided preconditions to <carry out emergency surgery>.

A striking feature of this event was the calling in of an external anaesthetist who lived near the hospital and who was acquainted with some of the physicians. This is in line with findings by Long, Cunningham, Carswell, and Braithwaite (2014), who concluded that both geographic proximity and past working relationships had significant effects on the choice of collaboration partners in health care.

Indeed, boundaries were crossed in this case entirely due to social relationships between professionals (e.g., obstetrician on duty and anaesthetist), rather than as a result of system design. This event also points to the importance of mapping the networks of professional and social relationships between caregivers. Professionals who are in high demand as a source of expert knowledge may need colleagues to act as slack sometimes. Creswick, Westbrook, and Braithwaite (2009) provide examples of what these networks look like in an emergency department in Australia.

Figure 13.1 also presents the average scores of the risks associated with variability sources, effectiveness of the Units of Slack and their respective standard deviations (SD). Based on these scores, it is possible to estimate an overall score for the gap between slack and risk. This gap was estimated by adding the SD to their respective risk scores, while subtracting the SD from their respective slack scores. Thus, the maximum overall risk was equal to: $38.6 + 38.6 + 17.5 + 27.3 + 45.6 + 11.3 = 178.9$. In turn, the minimum slack score was equal to: $64.4 + 76.9 + 83.4 + 80.1 + 56.6 + 61.5 = 422.9$. Therefore, the maximum gap (slack – risk) was equal to 244, which seems to make sense given that the outcome of the event was as successful as such events can be – i.e., the mother's life was saved. However, undesired outcomes in similar events are plausible due to non-linear interactions and functional resonance. 'Real' scores of risk and slack, in any given instantiation, may be affected by the specific nature of the couplings (e.g., timing and intensity), which in turn can either amplify or dampen the risk and slack of downstream functions.

It is also worth noting that, as the overall scores take into account the FRAM model as a whole, the definition of the system boundaries (i.e., which functions are in and out of the model) influences the analysis. When carrying out a retrospective analysis, as we did in this study, it is possible to define

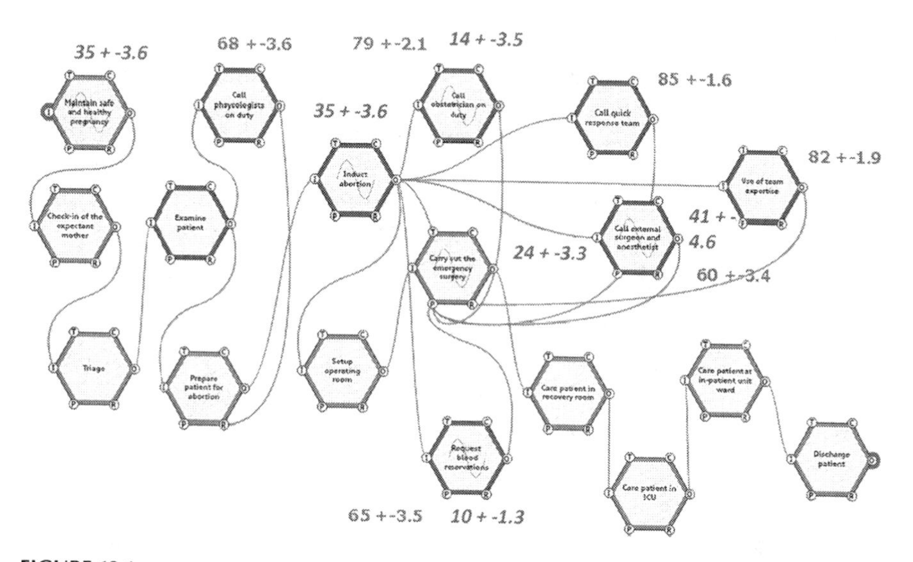

FIGURE 13.1
Instantiation of event 1. Notes: (i) waves inside hexagons indicate variability of the output; (ii) scores in italics represent the risks of the variability sources involved; and (iii) remaining scores represent the effectiveness of the Units of Slack.

the boundaries by including only functions that played an immediate role in the event, in terms of time and space, according to the data collected. This can be sufficient for identifying salient improvement opportunities as well as for having a reasonably easy-to-understand schematic representation of the FRAM model. Given the aforementioned limitations, the method used to calculate the gap, along with the interpretation, should be regarded as tentative, and could be further developed in future research.

The described event involved the collaborative use of slack across professional and departmental boundaries. Departmental boundaries were crossed when the blood bank was triggered, while professional boundaries were bridged when support from psychologists was requested, as well as when an external anaesthetist was called on, and the whole team of professionals carried out the emergency surgery.

It is also worth noting the implications of framing the outputs of four slack functions as preconditions for the surgery. This conveys that the slack with the longest lead time (i.e., time between the realisation of the need for slack and its actual availability at the point of use) determines the shortest possible lead time before activating the function that uses the slack. In this case, this critical slack corresponded to the external anaesthetist, who arrived at the hospital about 20 mins after being called at home. Slack that is a precondition could also be interpreted as part of the minimum standard resources that should be in place before initiating the function. If the function launches anyway, it can imply waste by 'making-do', which happens when a task begins without all its standard preconditions or a task continues to be executed even though at least one standard precondition is no longer available (Koskela, 2004). Saurin and Sanches (2016) draw a parallel between the concepts of making-do and resilience as follows: unsuccessful resilience actions are equivalent to making-do, implying waste or safety hazards; and successful resilience actions correspond to performance adjustment that tackles waste, without any undesired side effects.

Another insight from Figure 13.1 is that the use of most slack resources was triggered by the same function <induct abortion>. This suggests that the imprecise or late outputs of functions that directly involve patient care should be made visible as soon as possible to all concerned parties, so as the use of slack resource can be triggered. In this regard, the early detection of signs of deterioration of the patient clinical condition is key.

Lastly, the aforementioned classification of slack also proved useful for making sense of the described event. Table 13.2 presents the classification of the six deployed Units of Slack. All categorisations marked with an * indicate less than ideal situations, which can be interpreted as improvement opportunities. For instance, all Units of Slack received either moderate or low or scores regarding their visibility.

Event 2

Event 2 was also reported by an obstetrician. According to her report, a pregnant woman (36 weeks) sought the emergency department complaining that she had not been able to feel the foetus moving for the past 2 days. The obstetrician carried out an ultrasound exam, and the foetus's vital signs were fine: the heart rate was normal and the baby reacted normally to acoustic stimulation. However, the exam also revealed that the amniotic liquid was flowing more slowly than usual, which in turn could indicate that it was thicker than it should be. As an additional source of information to make the diagnosis, the attending physician made a call to the patient's private personal obstetrician, who was aware of the evolution of the pregnancy and overall health condition of the patient.

Based on the evidence gathered, the obstetrician decided to deliver the baby immediately. This would result in birth before the 37th week of pregnancy, which could be detrimental to the infant's health. Thus, there was a trade-off, since the decision to allow the pregnancy to continue could also put the baby at risk. Eventually, the woman underwent a caesarean section, which confirmed that the amniotic liquid was thick due to the presence of meconium (i.e., first evacuations of the baby); this reduced the oxygen available to the foetus. The newborn – whose condition was unstable – was immediately transferred to another hospital's neonatal ICU to receive treatment to lower the body temperature in order to preserve the baby's brain. This treatment was not available in the studied hospital. The mother was transferred to a recovery bed and then to a room, and was discharged 2 days

TABLE 13.2

Classification of the Units of Slack Deployed in Event 1

	Psychologists on Duty	Obstetricians on Duty	Quick Response Team	External Anaesthetist	Blood Bank	Team Expertise
Origin	Designed	Designed	Designed	Opportunistic*	Designed	Opportunistic
Nature of the resource	People	People	People and equipment	People	Material	People
Availability	Moderate*	High	High	Low*	Moderate*	High
Strategy of deployment	Standby resource	Standby resource	Standby resource	Standby resource, offline	Standby resource	Cognitive diversity
Visibility	Moderate*	Moderate*	Moderate*	Low*	Moderate*	Low*
Side-effects	Not identified	Not identified	Not identified	*Additional costs; physician unfamiliar with the patient	Not identified	Not identified
Durability	Not applicable	Not applicable	Members of this team need to go through a refresher course every 2 years	Not applicable	Expiration date	It tends to increase over time
Scope	Low	Moderate	Low	Moderate	Moderate	High
Is it a legal requirement?	No	Yes	Yes	No	Yes	No

later. The interviewed obstetrician reported that she kept in contact with the baby's family over the following months and was happy to know that the baby was growing up normally. Figure 13.2 presents the FRAM model of event 2, including the risk scores, effectiveness of the Units of Slack and the respective SDs. There were four slack functions in this event, namely: <call obstetrician on duty>, <use of obstetrician´s expertise>, <call private/personal obstetrician>, <transfer baby to neonatal ICU of another hospital>. In turn, these functions were triggered by three other functions: <triage>, <examine patient>, <care baby>.

Similarly to Event 1, an estimate was made of the overall score for the gap between slack and risk. Thus, the gap was equal to 76.3, which suggests that functions in this event were much more tightly coupled in comparison with event 1. A distinctive feature of event 2 was related to the collaborative use of slack across institutional boundaries. This occurred when it became necessary to transfer the newborn to the ICU of another hospital. This was facilitated by the existence of formal agreements between the studied hospital and other neighbouring hospitals, to make the transfer of neonatal ICU patients easier, especially in cases of overcrowding or infectious outbreak. Professional boundaries were also crossed when the emergency obstetrician requested support from the patient's private obstetrician, although he could not be physically present to help in the diagnosis.

Table 13.3 presents the classification of the four deployed Units of Slack. As with event 1, all categorisations marked with an * indicate improvement opportunities.

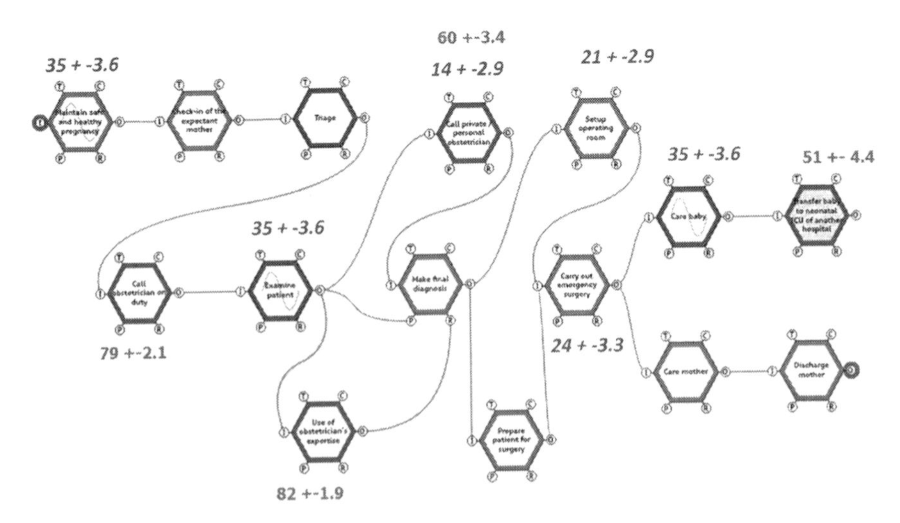

FIGURE 13.2
Instantiation of critical event 2. Notes: (i) waves inside hexagons indicate variability of the output; (ii) scores in italics represent the risks of the variability sources involved; and (iii) remaining scores represent the effectiveness of the Units of Slack.

TABLE 13.3

Classification of the Units of Slack Deployed in Event 2

	Obstetricians on Duty	Private/Personal Obstetrician	Neonatal ICU Beds in Other Hospitals	Team Expertise
Origin	Designed	Designed	Designed	Opportunistic
Nature of the resource	People	People	Equipment	People
Availability	High	High	Low*	High
Strategy of deployment	Standby resource	Standby resource	Standby resource, offline	Cognitive diversity
Visibility	Moderate*	Moderate*	Low*	Low*
Side-effects	Not identified	Not identified	*Additional costs to the hospital	Not identified
Durability	Not applicable	Not applicable	Not applicable	It tends to increase over time
Scope	Moderate	Moderate	Low	High
Is it a legal requirement?	Yes	Yes	No	No

Conclusions

This chapter discussed the use of slack in a maternity ward, emphasising its contribution to collaborative work across professional, departmental and institutional boundaries. Two case studies of deploying slack in obstetrics emergency care set an empirical basis for the discussion. A common feature of these two cases was their urgent nature, which probably made work across boundaries an imposition rather than an option. Also, the physical dispersion of some of the slack resources (e.g., centralised blood bank at the hospital, neonatal ICU beds in another hospital and external anaesthetist) implied that their use required work across boundaries in a physical sense, with the resulting time delays, which can be critical in an urgent situation. Thus, as a general design principle, setting aside financial criteria, slack resources should be located as close as possible to the point of use so that physical boundaries do not necessarily have to be crossed.

Additional findings to be emphasised are as follows: (i) the proposed categories for classifying slack shed light on the weaknesses and strengths that impact on collaborative work; (ii) the FRAM provides a model of the way slack resources are triggered and how they swarm to cope with variability and (iii) it seems possible to calculate a score measuring the gap between risk and protection for instantiations of FRAM models. This score may support redesign efforts, which could focus on either reducing risks or increasing the effectiveness of the slack.

References

Bardram, J. E. & Bossen, C. (2005, November). A Web of Coordinative Artifacts: Collaborative Work at a Hospital Ward. In *Proceedings of the 2005 International ACM SIGGROUP Conference on Supporting Group Work*, ACM, Sanibel Island, FL, pp. 168–176.

Betrán, A. P., Merialdi, M., Lauer, J. A., Bing-Shun, W., Thomas, J., Van Look, P., & Wagner, M. (2007). Rates of Caesarean Section: Analysis of Global, Regional and National Estimates. *Paediatric and Perinatal Epidemiology, 21*(2), 98–113.

Bourgeois III, L. J. (1981). On the Measurement of Organizational Slack. *Academy Management Review, 6*(1), 29–39.

Clarke, D. (2005). Human Redundancy in Complex, Hazardous Systems: A Theoretical Framework. *Safety Science, 43*(9), 655–677.

Creswick, N., Westbrook, J., & Braithwaite, J. (2009). Understanding Communication Networks in the Emergency Department. *BMC Health Services Research, 9*(1), 247. doi:10.1186/1472-6963-9-247

Greenhalgh, T. (2008). Role of Routines in Collaborative Work in Healthcare Organisations. *BMJ, 337*, a2448.

Hollnagel, E. (2012). *FRAM: The Functional Resonance Analysis Method: Modelling Complex Socio-Technical Systems*. London, UK: CRC Press.

Koskela, L. (2004, August). Making-do—The Eighth Category of Waste. *12th Annual Conference of the International Group for Lean Construction* (IGLC 12), Elsinor, Denmark.

Liker J. (2004). *The Toyota Way: 14 Management Principles from the World's Greatest Manufacturer*. New York, NY: McGraw-Hill.

Long, J., Cunningham, F., Carswell, P., & Braithwaite, J. (2014). Patterns of Collaboration in Complex Networks: The Example of a Translational Research Network. *BMC Health Services Research, 14*(1), 225.

Ong, M. & Coiera, E. (2010). Safety Through Redundancy: A Case Study of In-Hospital Patient Transfers. *Quality & Safety in Health Care, 19*(1), e32.

Perrow C. (1984). *Normal Accidents: Living with High-Risk Technologies*. Princeton, NJ: Princeton University Press.

Safayeni, F. & Purdy, L. (1991). A Behavioral Case Study of Just-in-Time Implementation. *Journal of Operations Management, 10* (2), 213–228.

Saurin, T. A. & Sanches, R. (2016). Making-do or Resilience: Making Sense of Variability. In F. Emuze & T. A. Saurin, (Eds.), *Value and Waste in Lean Construction* (pp. 15–22). London, UK: Routledge.

Saurin, T. A. & Werle, N. B. (2017). A Framework for the Analysis of Slack in Socio-Technical Systems. *Reliability Engineering and Systems Safety, 167*, 439–451.

Schöttle, A., Haghsheno, S., & Gehbauer, F. (2014, June). Defining Cooperation and Collaboration in the Context of Lean Construction. *Proceedings of 22nd Annual Conference of the International Group for Lean Construction*, Oslo, Norway, pp. 1269–1280.

Schulman, P. R. (1993). The Negotiated Order of Organizational Reliability. *Administration and Society, 5*(3), 353–372.

Sharfman, M., Wolf, G., Chase, R., & Tansik, D. (1998). Antecedents of Organizational Slack. *The Academy of Management Review, 13*(4), 601–614.

Silich, S., Wetz, R., Riebling, N., Coleman, C., Khoueiry, G., Rafeh, N., … Szerszen, A. (2011). Using Six Sigma Methodology to Reduce Patient Transfer Times from Floor to Critical-Care Beds. *Journal of Healthcare Quality, 34*(1), 44–54.

Stephens, R. J., Woods, D. D., Branlat, M., & Wears, R. L. (2011, June). Colliding Dilemmas: Interactions of Locally Adaptive Strategies in a Hospital Setting. *The Proceedings of the 4th Resilience Engineering Symposium*, Sophia Antipolis, France, pp. 256–262.

Wachs, P., Saurin, T. A., Righi, A., & Wears, R. (2016). Resilience Skills as Emergent Phenomena: A Study of Emergency Departments in Brazil and the United States. *Applied Ergonomics, 56*, 227–237.

14

Resilient Performance in Acute Health Care: Implementation of an Intervention across Care Boundaries

Robyn Clay-Williams and Brette Blakely
Macquarie University

Paul Lane, Siva Senthuran, and Andrew Johnson
Townsville Hospital and Health Service

CONTENTS

Introduction

This chapter will present the story of how an Australian Intensive Care Unit (ICU) in a large tertiary care hospital managed conflict between the ICU and surgery departments using principles of resilient thinking. This was not a poorly performing ICU in terms of administrative performance metrics, such as length of stay, mortality, bedside handover and trainee performance. The ICU did, however, have cross-boundary problems with the Department of Surgery. Elective surgery was frequently cancelled at short notice due to unavailability of ICU beds, resulting in poor interdepartmental relationships between surgery and ICU; this consequently fuelled conflict between clinicians within the ICU. Resilient thinking was used to develop and implement an ICU Escalation Plan (the Plan) to repair relationships between the two departments by establishing agreed rules for making ICU decisions around post-surgery beds.

The study hospital is unusual, in that it services a population of more than 750,000 spread over 750,000 km^2, and is 1,900 km from similar acute facilities that might otherwise be suitable for bypass in peak periods. This means there can be large and unexpected variation in requirement for the 14 adult ICU beds in the hospital; however, this must be managed within current resource constraints. To balance planning of major elective surgery with unpredictable emergency admissions, senior clinicians and managers in the hospital collaboratively developed the Plan to optimise flow of patients through ICU.

Description of the Intervention

The intervention consisted of an ICU Escalation Plan and introduction of a daily multidisciplinary morning meeting to determine ICU readiness state. The Plan (see Figure 14.1) is a workplace guideline with three readiness states:

- GREEN readiness indicates the ICU can accept a greater patient load, such that further elective cases could be considered.
- AMBER readiness indicates the system is approaching capacity in the next 24 hours such that elective surgery may need to be cancelled.

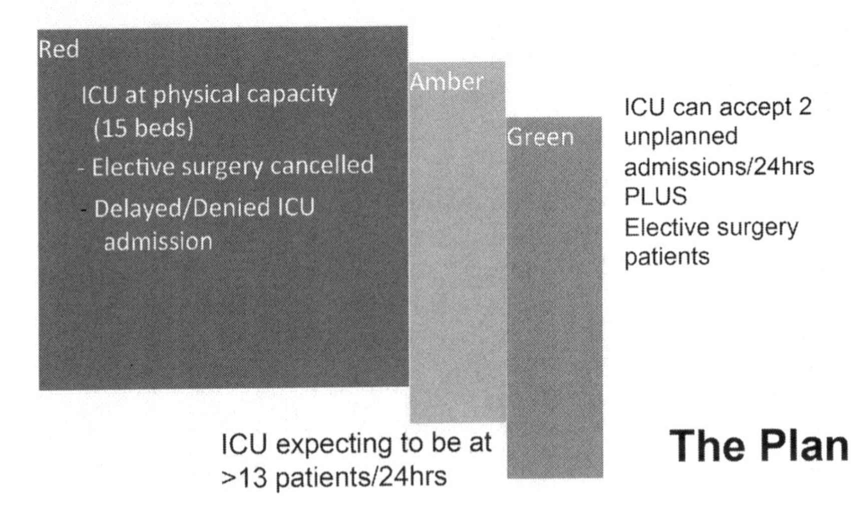

FIGURE 14.1
The ICU escalation plan.

- RED readiness indicates the ICU is at capacity and that surgery requiring ICU support will have to be cancelled and emergency admissions may not be able to be accommodated within normal resources.

Method

The pre-/post-intervention case study used a multi-method approach, consisting of process mapping, collection of audit data and two rounds of staff interviews. This chapter will present an analysis of the interview data, with a focus on boundary issues between clinicians within the ICU and between departments.

Interview participants were selected from ICU nurses, ICU specialist doctors and senior surgical and hospital management staff. Inductive interpretive analysis of transcribed interviews (Denzin & Lincoln, 2013) was undertaken to identify key themes. The first round of interviews aimed to collect staff perceptions of how ICU workplace cohesiveness varied with bed pressure and to collect data to develop a Functional Resonance Analysis Model (Clay-Williams, Hounsgaard, & Hollnagel, 2015) of the ICU bed availability decision-making process. The aim of the second round of interviews was to establish whether effects seen early in implementation persisted, changed or were lost, and to discover any new impressions after the natural evolution and bedding down of the plan following the morning meeting. To avoid bias, the second set of interviews were coded without reference to the first to see what themes were emergent.

Results

Twelve staff members were interviewed prior to implementation of the Plan, and 19 staff members were interviewed following the intervention. Perceptions of the utility of the Plan varied across the interviews from neutral, a position generally held by clinicians, to very positive, a position generally held by patient flow managers.

Round 1

The analysis of interview data revealed three themes: perceptions of the Plan, benefits of the Plan and processes associated with the morning meeting. Within each theme were a number of sub-themes. The Plan was perceived to have a number of functions: it made saying 'no' easier when the ICU was at capacity, it provided clear reference points for the concept of 'full' that were universal and not linked to bed numbers, it facilitated communication with managers about patient load and the need to transfer patients and it provided a useful basis for constructive conversations. It was also perceived, within the ICU, as providing agreement on the current state of readiness, and thereby offering more structure to decision-making processes. Despite this, it was recognised that the success of the Plan relied overly on the cooperation of those external to the ICU.

The Plan was also perceived as a 'canary in the coal mine' to identify system pressure. In this way, the status could be used as an indicator of proximal system operating point (Cook & Rasmussen, 2005) and provided a record and trend information on ICU performance and capacity. In particular, the Plan provided a *de facto* contract to drive behaviours for clinicians working across boundaries:

> ... it is meant to provide agreement across disciplines .... (Doctor 1)
> ... everyone works within this policy .... (Manager 4)
> ... there was almost like rules of engagement and people knew how decisions were made. (Manager 1)

Round 2

Some participants felt that the Plan was restricted to the ICU and lacked the authority to drive action in the rest of the hospital. They believed that bed block was a hospital-wide issue that went unaddressed, and that fixing the problem within the ICU was only a small part of the solution:

> I think there's no point having these things in place if the people at the other end aren't going to listen to you. There's no point saying when we get to code red we do this and this if no-one's going to listen to you when you get there. (AH/Nurse 5)
>
> They acknowledged that they'd received a call, but not a lot happened with them, so there's still—the service group still expects it's going to be managed by us. We're going to manage the surgeons and everything else as well as managing ICU. They still haven't got it in the service group. (Manager 2)

There was also some residual scepticism over the functioning of other departments and the effect of the political environment on their management. While Emergency Department (ED) and elective surgery targets were often cited as potentially influencing referrals, it was also suggested that the visibility of the ICU could be further improved:

> Then I think the other thing is, not so much transparency because that's what everybody talks about, but more visibility so that we can understand their challenges and constraints. We're not there to fix them, but also so that they can understand ours. Because sometimes it feels like,

> when you're in the Emergency Department, for example, you're the fish bowl that everybody can look at, but we can't see what anybody else is doing, which is a chip on the Emergency guys' shoulders sometimes—which we also need to drop. But I think it's nice to see the other person's pressures as well. (Doctor 7)

Others who were neutral about the effectiveness of the Plan cited other changes in the ICU and surgery as the reasons for improvements in workflow. One suggestion was that the ICU had more bed availability due to improved staffing:

> I see merit in it, but whether it's been useful? Maybe a little bit. Maybe it's, I don't know, maybe it's probably not needed as much now that they're resourced a bit better, nursing-wise. [Facilitator: Okay. So, the resources and other factors matter more critically than ...] "Yeah, definitely." (AH/Nurse 4)

It was also suggested that a change to the surgical diary, which spread out the workload, had contributed to the reduced cancellations:

> But I suppose if I compare it to what it was 4 years ago, 5 years ago, it certainly is improved a little bit and I think the biggest factor is that having that fifth day at theatre .... (Doctor 6)

An emergent theme from Round 2 was the increased ability of participants to approach problems from a system perspective, bypassing the usual 'silo'-based view. In fact, the second and third major themes that were apparent throughout the interviews were patient flow and hospital-wide context. Some participants discussed the pressures other teams experienced, demonstrating an improvement in hospital-wide understanding and cohesion. This is particularly relevant given the context of National Performance Targets, the National Emergency Access Target and the National Elective Surgery Target. These targets apply competing pressure on beds through a requirement to, respectively, clear patients through the ED within 4 hours of arrival and perform planned surgeries within predetermined time frames:

> I guess it's a lot of things. We have departments that have guidelines like ED will have a four-hour guideline to get a patient to a ward. So, then they want to push a patient to you because that's their guideline. Rightly so, they're trying to do their job. (AH/Nurse 5)

Some cohesion seemed to stem from a practitioner versus political management divide, which focused the issues on system pressure and constraints, rather than individuals within departments:

> The relationship between the ICU and ED here is often fraught. I think in recent years since the advent of the four-hour rule has become far more difficult. They're under a singular set of pressures that really at one time they would have intervened and seen what the results of interventions were, whereas now patients can still be relatively half baked in terms of

> management. Things are still in rapid evolution but the four-hour mark is coming up and so they'll make a referral. We all sometimes see as being frivolous. Of course, they've got different objectives, which are sometimes not patient-centred objectives, but I think everyone recognises that. (Doctor 2)
>
> Then for the elective stuff, as I say there's a lot of pressure from the NES targets to still get everything done regardless. That's why I think sometimes—and this maybe me being cynical—there's a little bit of political game playing in terms of how they—or what goes on the theatre list …. (Doctor 3)

Overall, those who felt the Plan was not able to resolve patient flow issues tended to see the problems as a result of systems or management errors, but animosity towards particular people or units was reduced.

In general, patient flow managers felt more optimistic about the Plan than front-line clinicians. However, participants felt that by providing agreement on current status, the Plan gave more structure to decision-making processes within the ICU:

> I think by reducing the ad hoc nature of the decisions that makes it clearer. I think any, you know, the old 'good fences make good neighbours'. I think it helps from that perspective. I think it probably has improved our workflow. Not so much the morning meeting but the people having an idea about our bed state has improved our workflow to some degree and that helps—then they can say yes we're going to go ahead with all the surgery or we're going to can all the surgery. …. We had in the past where individual surgeons would come marching up and say, 'Well, I want to do my case'. That's gone away, which is a very, very good thing. (Doctor 2)

At 7 months post-implementation, the key improvements in cohesion and communication found during Round 1 were further reinforced. Analysis of coding showed that communication was the most discussed theme, with a focus on personal communication and relationships.

The improvement in the ICU's internal workflow and communication was seen as going some way to improving the clarity and visibility of the ICU and flowed on to increasing cooperation with other department and hospital management:

> Basically, what I can say to you is it's communication between the nursing director and the wards that we transfer to. It's just that network we've built up. We've realised the importance of it. It's the traffic light system that's actually helped us see that. When they see that we're at this and we don't have a lot of room to move, they will support us in taking the patients out, rather than bed blocking. (Manager 3)

Interpersonal communication was a major theme in the Round 2 interviews. Emphasis was placed on direct communication and increased familiarity between people:

> Then, whereas, if they know each other then—you know what it's like, you know anybody then you're more inclined to be, to facilitate things, rather than be—not obstructive, but not be as helpful. (Doctor 7)

In conjunction, team cohesion was seen to emerge from the daily team meetings. The Plan was used to create a unified mental model and agreed processes underpinning plans for daily action:

> I think bringing the whole team together and everyone hearing the same thing, and knowing what elective surgery are and knowing what our bed capacity is—I think is a very useful thing. I think it's been good to incorporate nursing and allied health into that, as well. Just so everyone is on the same page, and in terms of a team building exercise. (Doctor 1)
>
> So, we have lots of people—like the social worker comes, the speech pathologist—I think that's great. Everyone's on the same page. We never used to have that before. (AH/Nurse 3)

Interestingly, this process of coming to agreement on a bed status each morning could be seen as a team building exercise in itself. The ICU now start every day with a team negotiation that brings everyone together to discuss the treatment and longer-term planning for each ICU patient. At the end of the meeting, a consensus is reached on the ICU status for the day in accordance with the Plan. The unified ICU status then provides a foundation for all other conversations and interaction within the team for the rest of the day, and also presents a unified voice for the team when communicating with external departments:

> I guess it's more of a team environment, multi-disciplinary. I think that's better for the patients we look after. So, there's more of a team approach. I think communication's a lot better. Everyone seems to be on the same page more. (AH/Nurse 4)
>
> In this unit alone, we have a joint morning meeting at 8 o'clock in the morning. That's probably one of the biggest changes that's come into effect in the last year I'd say within the unit, over the 15 years I've been here. Mainly because everyone's involved, everyone knows what's happening. I think by doing that, everyone's more confident with each other. That comes down then if things happen in the unit you can rely on people and you know who they are, and you know what their skills and qualities and that are too. (AH/Nurse 5)

The opportunity to raise concerns within a structured meeting environment gives everyone a voice, thereby reducing frustration that might lead to instances of aggravated conflict between individuals. The creation of a single team mental model influences interaction externally as well as communicating a sense of clear ownership and accountability:

> No, overall I'd say that the ICU is working well. I think they're a really cohesive team. I think the steps they've taken to try and manage that uncertainty, that being a positive thing. I think the actual putting something in place that people can own has helped with the relationship in the team, that's great. I think a lot of this is also around the difficulties

> of you could get the different decision depending on who was there. So, having something they could all own and that people recognise this is how we manage and that the other services understand that, that helps. So, I think that's certainly, I'd say they were a cohesive, well functioning team. Yes, there's pressure but they manage it well. (Manager 1)

Discussion

We found that both organisational and situational boundaries existed in the hospital system. The physical ward-based structure of hospitals and the organisational arrangement of hospitals into departments encourage physical and organisational boundaries to integrated care. Professional boundaries (Powell & Davies, 2012) in the study were reinforced by the fact that it is a 'closed ICU', whereby the authority for patient care is transferred to an intensivist for the duration of the patient's stay, rather than remaining with the patient's normal surgeon or physician. In our data, references to boundaries were not explicit, but could be seen in references to physical, professional group, role, conceptual and personality-based boundaries. Organisational boundaries can arise from the way hospitals are designed and built, from the way departments are structured and resourced (e.g., closed ICU and ED *'fish bowl'* [Doctor 7]), from the appointment of designated leaders (e.g., department directors, service group) and from guidelines that drive the way care is delivered (e.g., *'four hour rule'* [Doctor 2]). Many hospital quality management and clinical governance processes, such as appointment and rostering of staff, implementation of care guidelines and accreditation, reinforce these organisational boundaries.

In contrast, situational boundaries can arise from the way individuals practice (e.g., *'individual surgeons would come marching up'* [Doctor 2]), from the way teams are structured, as a 'pushback' against pressure from managers or other departments (e.g., *'the service group still expects it's going to be managed by us'* [Manager 2]), and from the varying needs of individual patients and their families as they pass through the system. The closed ICU arrangement has the potential to amplify daily differences between Work-as-Imagined (and care-as-desired) by surgeons/physicians and Work-as-Done by intensivists. We found some 'boundary spanners' (Long, Cunningham, & Braithwaite, 2013) in terms of surgeons who visited the ICU and intensivists who collaborated with their surgical colleagues, but most clinicians remained relatively isolated within their clinical departments. While situational boundaries would be seemingly easier to span than organisational boundaries, often these boundaries are emergent and reinforced by cultural norms, and thereby extremely resistant to intervention.

The Plan had the capacity to function as either a device for overcoming boundaries or an enforcer of boundaries, depending on who was wielding the tool. By providing a structured way of communicating between departments, for example, the Plan was successful in bridging some of the natural hospital boundaries. In contrast, the Plan also provided a means to say 'no' to demands from other departments and to protect the ICU as a bounded 'space' (e.g., *'good fences make good neighbours'* [Doctor 2]). In this case, the boundary, while considered 'negative' by the surgical department, could actually be considered 'positive' by the ICU.

As expected, Work-as-Imagined and Work-as-Done varied among those involved with ICU patient flow. The Plan impacted perception and understanding of patient flow processes, contributing to better communication and teamwork between the ICU and surgery departments. Our study showed how a device as simple as an escalation instruction could facilitate improved working across department boundaries. The 'boundary spanners' (Long et al., 2013), in particular, appeared to gain the most from the intervention. 'Boundary spanners' are individuals within organisations who understand contextual information on both sides of the boundary sufficiently well to be able to act as go-betweens, enabling them to pass information between groups (Tushman & Scanlan, 1981). In our study, the 'boundary spanners' consisted of clinicians and managers within the hospital, whose formal roles involved negotiation between departments, such as the ICU Nurse Unit Manager and the hospital patient flow manager, and those who socialised across groups, such as some of the senior ICU doctors. As these individuals were the 'face' of their departments when negotiating bed availability or fielding external complaints, they were the ones most likely to have the tough conversations and be most affected by the previously poor interdepartmental relationships. It was therefore an important finding that these individuals benefitted from the intervention.

By establishing rules for decision-making around ICU bed allocation, the intervention improved internal professional relationships within the ICU and between the ICU and external departments. The reduced rate of elective surgery cancellations that were found to follow the intervention reflects a more resilient system. It is interesting that offering extra ICU capacity to the surgeons when the ICU was in a state of GREEN, while not practically useful (due to the lead time involved in scheduling additional surgical patients), accumulated a store of goodwill on a social level. This store of goodwill appeared to strengthen system resilience by providing 'boundary spanners' with resources to draw on to provide a better foundation for conversations in the event that pressure increases.

Conclusion

The methodology adopted for this evaluation allows behaviour to be researched along with quantitative outcome data. Interviews with staff both early in implementation and after the Plan had been in operation for several months were able to capture the mechanisms by which the Plan improves workflow within the ICU and relations with other departments. They also provided a deeper understanding of why interventions worked. Analysis of the audit data alone would have missed these mechanisms of action.

This study also revealed a complementary effect of the Plan and associated meetings that, in addition to fixing a workflow issue and improving communication, strengthened cohesion through functioning as an ongoing team building exercise. As such it was able to become stronger over time, rather than dissipate with changes in staff, and make the culture itself more resilient within the unique environment of health care. The unit-level intervention aided negotiation across organisational boundaries, which was of particular benefit to the essential work of the departmental 'boundary spanners'.

References

Clay-Williams, R., Hounsgaard, J., & Hollnagel, E. (2015). Where the Rubber Meets the Road: Using FRAM to Align Work-as-Imagined with Work-as-Done When Implementing Clinical Guidelines. *Implementation Science, 10*(1), 125.

Cook, R. & Rasmussen, J. (2005). "Going Solid": A Model of System Dynamics and Consequences for Patient Safety. *Quality & Safety in Health Care, 14*(2), 130.

Denzin, N. K. & Lincoln, Y. S. (2013). *Strategies of Qualitative Inquiry*, 4th ed. Thousand Oaks, CA: Sage Publications.

Long, J. C., Cunningham, F. C., & Braithwaite, J. (2013). Bridges, Brokers and Boundary Spanners in Collaborative Networks: A Systematic Review. *BMC Health Services Research, 13*(1), 158.

Powell, A. E. & Davies, H. T. (2012). The Struggle to Improve Patient Care in the Face of Professional Boundaries. *Social Science & Medicine, 75*(5), 807–814.

Tushman, M. L. & Scanlan, T. J. (1981). Boundary Spanning Individuals: Their Role in Information Transfer and Their Antecedents. *Academy of Management Journal, 24*(2), 289–305.

Part V

Closure

15

Discussion, Integration and Concluding Remarks

Jeffrey Braithwaite
Macquarie University

Erik Hollnagel
Jönköping University

Garth S. Hunte
University of British Columbia

Across the pages of the book our authors have grappled with boundary actions, concepts, theories and structures, seeking to understand and elucidate the relationship between resilient performance as expressed in health care settings and the boundaries that are an inevitable part of how work is done and organised. By way of beginning, in the section entitled *Openings*, Braithwaite, Hollnagel and Hunte provided a brief historical context – encapsulating the essence of the first four books – to acknowledge the journey to date. Braithwaite, Hollnagel and the late Bob Wears used a sailing metaphor in Chapter 2 to add some texture to the boundary ideas that followed in the subsequent chapters.

Turning to the two chapters which constituted Part 2 of the book, *Negotiating Across Boundaries*, both Johnson, Lane, Klug and Clay-Williams in Chapter 3 and then Robson in Chapter 4 made a case for understanding resilient performance by showing the centrality of formal and informal negotiating in complex health care settings. Negotiating always happens across multiple boundaries: those between the partners to the negotiation, as well as the slippery conceptual, ideological, political, sociological and psychological barriers that can separate negotiators and their relative stances.

In the third section, *Theorising About Boundaries*, Sheps and Wears' contribution was to navigate resilience theory, a particularly thorny problem for those in the Resilient Health Care Network. How does resilience theory apply in the real world, they asked? Zhuravsky, Lofquist and Braithwaite, by way of contrast, showed that leading health care organisations and services require distributed, diffused, shared leadership models: gap-bridging, boundary-crossing leadership, if you will. Patterson, Dieckmann and Deutsch looked at another area where the gap between theory and practice can be fraught, exploring the boundaries between simulation and clinical work and the ways in which simulation can work to break down barriers.

Having reflected on resilience and boundary concepts and theories in Parts 1–3, we now had a platform for the next, and largest, section: Part 4, *Empiricising Boundaries.* Churruca, Long, Ellis and Braithwaite provided a context for what followed through a review of 30 cases drawn from chapters in the four previous volumes of RHC, teasing out various boundary actions across the case studies. Dieckmann, Clemmensen and Lahlou furnished a video-ethnographic description to close the gap between real and imagined boundaries in the world of medication dispensing. Next, Hegde and Jackson changed tack, taking a close look at the role of operating room floor managers: a classic boundary-spanning role, being the glue of effective care in a busy, longitudinal, emergent workplace. Riffing further on the theme of patients and their flow through caring systems and the resultant cross-boundary actions needed to make the system work, Back, Ross, Jaye, Henderson and Anderson examined transitions of National Health Service patients across organisational boundaries. Staying with the UK, Sujan, Huang and Biggerstaff looked at trust in relation to setting, observing and respecting boundaries from the inside as well as the outside. Changing countries and settings, Werle, Saurin and Soliman immersed themselves in a maternity ward, looking at a boundary category not dealt with in other chapters: the phenomenon of slack or the buffering and redundancy which is (and needs to be) built into every functioning system. Penultimately, Clay-Williams, Lane, Blakely, Senthuran and Johnson probed a health systems intervention, examining the effects of the implementation of an ICU escalation plan as a device for building deep knowledge of the system that the intervention perturbed.

Finally, in Part 5, *Closure,* we provide this concluding chapter. Table 15.1 summarises the outcomes of the book, with selected key lessons from the authors' countries, the empirical stances they took and the theoretical approaches they embraced.

At the conclusion of our discursive, theoretical and empirical examination of the various mechanisms, models and approaches to working across boundaries, we end by offering some practical advice. Boundaries are omnipresent and intrinsically interesting. They can act as signals to those who apprehend them or are looking for them. They tell us something about where one thing (a category, some behaviour, a concept or a social structure, say) ends and another begins. They can shine a light on where the structural holes in systems are.

TABLE 15.1

A Summary of the Book – Authors, Lessons, Country, Empirical Stance and Theoretical Approach

Authors, Chapter	Selected Key Lessons	Country	Empirical Stance	Theoretical Approach
Part 1: Openings				
Braithwaite, Hollnagel and Hunte Introduction: The Journey to Here and What Happens Next	Prior books in the series focused on resilience and everyday clinical work, Work-as-Imagined and Work-as-Done, and research and theory on ways to deliver resilient health care. Now, we discuss resilient health care in the context of complexity of social structures and different kinds of boundary-crossing actions.	Australia, Denmark, Canada	Historical background and context of Volumes 1–4	Introduction and conceptual overview
Braithwaite, Hollnagel and Wears Bon Voyage: Navigating the Boundaries of Resilient Health Care	On the nature of boundaries, gaps and bridges: ways in which to work across boundaries to enable resilient performance across systems, organisations and services. Applying the metaphor of sailing to frame what follows.	Australia, Denmark, United States of America	Explores the ideas of boundaries, boundary-crossing and the relationship to resilient health care. Establishes the concept of boundaries and encapsulates the idea to frame the rest of the book	Key theoretical ideas about boundaries, gaps and bridges
Part 2: Negotiating Across Boundaries				
Johnson, Lane, Klug and Clay-Williams Working across Boundaries: Creating Value and Producing Safety in Health Care Using Empathic Negotiation Skills	Explores the benefits of interest-based bargaining and the application of negotiation skills; examines styles of negotiation in a five-fold model comprising the Collaborator; the Competitor; the Avoider; the Accommodator; and the Compromiser.	Australia	Case study applying these ideas to health care	Negotiating across boundaries

(*Continued*)

TABLE 15.1 (*Continued*)

A Summary of the Book – Authors, Lessons, Country, Empirical Stance and Theoretical Approach

Authors, Chapter	Selected Key Lessons	Country	Empirical Stance	Theoretical Approach
Robson Untangling Conflict in Health Care	Perspectives on engaging with conflict and negotiating across sub-systems. Includes a differentiation of power-based approaches, rules approaches and interest approaches, and advocates for a relational/narrative approach for engaging with natural, everyday conflict in complex systems.	Canada	Case study of negotiations in complex adaptive systems	Complexity, conflict, collaboration, relationality
Part 3: Theorising About Boundaries				
Sheps and Wears 'Practical' Resilience: Misapplication of Theory?	An examination of some historical and current themes in patient safety and resilient health care.	Canada, United States of America	A theoretical survey of a range of ideas around resilience	Resilience as a deeply embedded potentiality
Zhuravsky, Lofquist and Braithwaite Creating Resilience in Health Care Organisations through Various Forms of Shared Leadership	Shared leadership, acting to mitigate WAI-WAD via boundary-crossing concepts.	New Zealand, Norway, Australia	Theoretical in nature primarily, with case examples	Shared leadership as a paradigm for dealing with boundary actions
Patterson, Dieckmann and Deutsch Simulation: A Tool to Detect and Traverse Boundaries	Discursive and theoretical contribution of simulation in different modes to break down barriers. The relationship between simulation to clinical work, in particular, to facilitate adaptive expertise, and resilient performance is explored.	United States of America, Denmark	Simulation and resilience	Theorisation of simulation in situ and outside the workplace

(*Continued*)

TABLE 15.1 (*Continued*)

A Summary of the Book – Authors, Lessons, Country, Empirical Stance and Theoretical Approach

Authors, Chapter	Selected Key Lessons	Country	Empirical Stance	Theoretical Approach
Part 4: Empiricising Boundaries				
Churruca, Long, Ellis and Braithwaite Looking Back over the Boundaries in Our Systems and Knowledge	A review of 30 cases drawn from chapters in the four previous volumes of Resilient Health Care. Teases out boundary actions across the case studies.	Australia	Aggregation of empirical cases, probing activities which enabled resilient performance through working across boundaries	Empirical examination of case study theory across multiple RHC volumes
Dieckmann, Clemmensen and Lahlou Understanding Medication Dispensing as Done in Real Work Settings – Combining Conceptual Models and an Empirical Approach	An analysis of medical processes in real-world medication dispensing settings, using video ethnography. Subject-centred video ethnography methods allow a perspective in real-world settings.	Denmark	Contribution to theory, linking installation theory with the Functional Resonance Analysis Method and an empirical case study videoing drug dispensation	Combining conceptual models. Juxtaposing installation theory and the FRAM approach
Hegde and Jackson Resilient Front-Line Management of the Operating Room Floor: The Role of Boundaries and Coordination	Analysis of resilience on the front-lines from the vantage point of one key player – the floor manager. The floor manager can effectively coordinate across physical, functional and hierarchical boundaries.	United States of America	Empirical study of the work, communication, expressions and interactions of 15 floor managers in the Operating Room	Capacity for Manoeuvring (CFM) and the floor managers' role as a bridge to effective systems performance
Back, Ross, Jaye, Henderson and Anderson Patient Flow Management: Codified and Opportunistic Escalation Actions	Patient flow through escalation protocols, and referrals and transitions of patients in a hospital setting are examined. Patient flow traverses boundaries and silos, be they clinical, technical or otherwise.	United Kingdom	Empirical case studies across the NHS and a conceptualization of when demand exceeds capacity	Escalation and patient flow management. Watchstander role as key meso-level action to facilitate boundary negotiation.

(*Continued*)

TABLE 15.1 (*Continued*)

A Summary of the Book – Authors, Lessons, Country, Empirical Stance and Theoretical Approach

Authors, Chapter	Selected Key Lessons	Country	Empirical Stance	Theoretical Approach
Sujan, Huang and Biggerstaff Trust and Psychological Safety as Facilitators of Resilient Health Care	A study of trust and psychological safety in the context of dynamic trade-offs, and their relationship to resilient health care	United Kingdom	Empirical case studies of patient transfers from Emergency Departments to wards and discharge of patients on a warfarin regime	Trust and psychological safety in emergency departments and amongst discharged patients
Werle, Suarin and Soliman Collaborative Use of Slack Resources as a Support to Resilience: Study of a Maternity Ward	Theoretical analysis of slack and empirical examination of two events using interviews and FRAM of each event.	Brazil	Case study classification of slack resources, applied to a maternity ward via FRAM analyses	Slack resources and resilience in maternity wards
Clay-Williams, Lane, Blakely, Senthuran and Johnson Resilient Performance in Acute Health Care: Implementation of an Intervention across Care Boundaries	Examination of how an intervention crosses boundaries in care settings (i.e., an intensive care unit). Well-intentioned boundaries can become barriers over time; discusses WAI–WAD aspects of an intervention.	Australia	Empirical study of escalation of intensive care using multi-methods in Townsville, Australia	Resilient performance in intensive care; conflict between intensive care and surgical departments
Part 5: Closure				
Braithwaite, Hollnagel and Hunte Discussion, Integration and Concluding Remarks	Working the boundaries, working with boundaries and working across boundaries.	Australia, Denmark, Canada	Brings the major empirical themes together	Finalises the book and discusses the benefits to different stakeholders

Indeed, as Table 15.1 shows, boundaries at bottom offer an opportunity for exploitation (the *tertius gaudens* operators amongst us) or for joining things together (the *tertius iungens* approach). They can stimulate us to action or demotivate us by telling us how big the gulf is between where we are now and how much we have to do to get across it.

When thinking about this, we have reflected on what the attributes are of tertius gaudens and tertius iungens that contribute or detract from the good? These examples were provided in the introduction as meso-level gap bridging actors, but gaudens was depicted as someone who plays politics and exploits the role for their own benefit (through divide and conquer or backstabbing strategies). Such an approach to power is manipulative and unethical. Is this the opportunity for exploitation that boundaries offer us, and if so, how does that contribute to resilient performance?

Ambidexterity and balance between exploration and exploitation may improve performance, but this is not the sense of exploitation depicted in tertius gaudens. Multiple studies highlight the importance of building ambidextrous capabilities to achieve both exploitation of current knowledge and technologies for present performance, and exploration of new knowledge and technologies to adapt to and prepare for future demands. The duality of this challenge is often a dilemma.

It seems to us that the third who joins (iungens) is more likely to contribute to and facilitate resilient performance than the third who benefits (gaudens). Hence, we emphasize the iungens strategic orientation, which endorses connecting organizations and emphasizes the collective good, as befitting the watch stander, broker, and navigator.

After digesting the book across the landscape of its chapters, our view, and we hope yours, is thus: for resilient health care performance, boundaries cannot be ignored. They are to be mapped, traversed, exploited and navigated. In the end, one of our desires is to support resilient performance wherever it is found – wherever it is expressed. That's a desire we share with those who aren't as immersed in resilient health care as we are, but who seek better, more integrated and more collaborative structures through which care is delivered, policy is developed and enacted, and services are led and managed. In practice this means all of us – health care politicians, policy makers, researchers, managers, leaders, clinical staff, support workers and patients, their friends, relatives and carers. That means, we hope, that in a book like this, there's something for everyone.

Index

E

F

G

H

I

L

M

N

O

P

R

S